CONSTRUCTION SURVEYING, LAYOUT, AND DIMENSION CONTROL

CONSTRUCTION SURVEYING, LAYOUT, AND DIMENSION CONTROL

Jack Roberts, Ph.D.

Delmar Publishers Inc.™

I**(T)**P™ An International Thomson Publishing Company

New York • London • Bonn • Boston • Detroit • Madrid • Melbourne • Mexico City • Paris
Singapore • Tokyo • Toronto • Washington • Albany NY • Belmont CA • Cincinnati OH

NOTICE TO THE READER

Publisher does not warrant or guarantee any of the products described herein or perform any independent analysis in connection with any of the product information contained herein. Publisher does not assume, and expressly disclaims, any obligation to obtain and include information other than that provided to it by the manufacturer.

The reader is expressly warned to consider and adopt all safety precautions that might be indicated by the activities described herein and to avoid all potential hazards. By following the instructions contained herein, the reader willingly assumes all risks in connection with such instructions.

The publisher makes no representations or warranties of any kind, including but not limited to, the warranties of fitness for particular purpose or merchantability, nor are any such representations implied with respect to the material set forth herein, and the publisher takes no responsibility with respect to such material. The publisher shall not be liable for any special, consequential or exemplary damages resulting, in whole or in part, from readers' use of, or reliance upon, this material.

COPYRIGHT © 1995
By Delmar Publishers Inc.
an International Thomson Publishing Company

I(T)P˚ The ITP logo is a trademark under license.

Printed in the United States of America

For more information, contact:
Delmar Publishers Inc.
3 Columbia Circle, Box 15015
Albany, New York 12203-5015

International Thomson Publishing
Berkshire House
168–173 High Holborn
London, WC1V7AA
England

Thomas Nelson Australia
102 Dodds Street
South Melbourne 3205
Victoria, Australia

Nelson Canada
1120 Birchmont Road
Scarborough, Ontario
M1K 5G4, Canada

Cover Design: Charles Cummings Advertising/Art, Inc.

Delmar Staff
Associate Editor: Kimberly Davies
Development Editor: Jeanne Mesick
Editorial Assistant: Donna Leto
Project Editor: Patricia Konczeski
Production Coordinator: Karen Smith
Art/Design Coordinator: Cheri Plasse

International Thomson Publishing GmbH
Konigswinterer Str. 418
53227 Bonn
Germany

International Thomson Publishing Asia
221 Henderson Bldg. #05–10
Singapore 0315

International Thomson Publishing Japan
Kyowa Building, 3F
2-2-1 Hirakawa-cho
Chiyoda-ku, Tokyo 102
Japan

1 2 3 4 5 6 7 8 9 10 XXX 01 00 99 98 97 96 95

Library of Congress Cataloging-in-Publication Data

Roberts, Jack, 1942–
    Construction surveying, layout, and dimension control / Jack Roberts
        p.   cm.
    Includes index.
    ISBN: 0-8273-5723-0
    1. Surveying. 2. Building sites. 3. Building layout I. Title.
TA625.R63 1994
624—dc20   94-28591
                                                    CIP

CONTENTS

PREFACE

Construction is like any other skill area in today's world; it has become more complicated and more specialized. The days when one person "ran the job" from start to finish are almost gone, even on the smaller projects. This age of specialization has produced an entire cadre of persons whose skills must be utilized and coordinated to effectively complete a construction project.

The processes of gathering information about a proposed job site, laying out the location of structures on the site, checking the dimensions of the structures during construction, and documenting the completed work is now considered to be a distinct, specialized, and separate skill area that requires specialized training. Most colleges and university construction education programs now provide such training as an integral part of their construction curriculum. Many trade organizations and apprentice programs offer similar training through local continuing education classes.

One of the problems associated with providing specialized training in any field is the lack of texts and other reference material dedicated to that field. The instructor simply does without or chooses a text from a related discipline, which he or she then must adapt to the subject under study. Traditionally, surveying texts have been chosen and adapted for use in construction layout and control classes. However, the texts that work so well in studying the science of surveying often do not work at all well in the construction discipline. Much of the content of the surveying text is of no relevance to the construction professional or is too theoretical to be applicable.

The primary purpose of this text is to provide a basic reference work for the construction student who is taking a course in which construction project layout and control is the primary focus of the course.

This text is divided into two parts. The first part is dedicated to developing the basics of surveying skills while referencing these skills to construction activities. The second part of the text is dedicated to applying the skills learned in the first part to actual construction processes. The processes are presented in the chronological order of construction. Additional chapters on developing as-built documents and space planning are included.

The secondary purpose of this text is to provide a ready reference for those who need to refresh their memories in selected areas of construction project layout and control techniques. An appendix is included which contains many of the mensuration tables and data used in construction offices and on job sites on a daily basis.

The step-by-step procedures, checklists, and notes to the reader contained in the text are based on the actual experiences of the author and other practicing construction professionals. Every effort has been made to make this text a "real life" and "down to earth" publication. The chapter questions and exercises are designed to reinforce pertinent data in the chapter. The instructor's guide contains information and suggestions on classroom, laboratory, and field activities which have been proven to work. Rainy day activities are noted.

A note about the terminology used in this text: Due to the specialized nature of measurement and control activities as used in construction work, the person who does this type of work has been given a special job title by many companies and on many job sites—"Field Engineer."

The title of "Field Engineer" does not imply that the practitioner is a graduate of an engineering school, nor does it assume any type of professional registration. Rather, it is used in the same vein as the title of "Locomotive Engineer." It simply means someone who uses technical equipment to accomplish a technically oriented task.

When using the term "Field Engineer," the author intends no distinction as to gender. In cases where the pronouns *he* or *she* were required, both have been used to ensure that no gender-bias language occurs.

The terminology used in the chapters dealing with curves was chosen after considerable research. The CT, TC, PI, BVC, EVC, and other notations seem to be the most widely used terms in this discipline. The author apologizes to the instructor or practitioner who is accustomed to other terminology and sincerely hopes that a conversion can be made without undue difficulty.

The author offers this text with the expectation that it will be a "keeper" and that it will find its way to the reference shelf of many construction professionals.

ACKNOWLEDGMENTS

The author wishes to acknowledge the aid, advice, and assistance received from the following individuals: Rick Sward, Buddy Bell, Cary Hughes, Berkman Manuel, Phil Freeland, Art Daniel, Monty Howard, Julie Dunnahoe, Ed Trejo, Mark Pfeiler, Mike Cheshier, Lisa Brewster, Steve Hendricks, Mark Barry, Ben Pool, and Derek Nalls.

The following organizations and companies have provided invaluable input and material for this text: Harrison, Walker, and Harper Construction, Granite Construction, Austin Commercial Inc., Spectra Physics, Lietz Sokkia Corp., Cooper Tools, Boring and Tunneling Company of America, Griswold Machine and Engineering, and Barry, Kneeland, and Schlatter, Pfeiler & Associates Engineers, and the International Union of Operating Engineers.

The author and Delmar Publishers wish to thank the following reviewers for their valuable contributions:

R. S. Davenport
Sand Hills Community College

John Erion
Bowling Green State University and Clark State Community College

Ron Gallagher
University of Toledo

Richard Haley
Rock Valley College

Charles Matthewson
Southern Illinois University

Carl Sellers
SUNY, Morrisville

Steve Williams
Auburn University

As in all projects such as this, there are many individuals and organizations who assist in so many ways. Of course, they are too many to name, but their assistance is appreciated and acknowledged.

A special thanks to the staff at Delmar Publishers, Inc. Without their guidance and encouragement, this project would never have seen the light of day.

Lastly, thanks to the family for bearing with me while I took time from you to complete this project.

I

BASIC SURVEYING PRACTICES

1

INTRODUCTION

1-1   History of Surveying

Evidence of land measurement, marking, and the associated record keeping occurs in some of the earliest recorded histories of civilization. Proverbs 22:28 in the Old Testament portion of the Christian Bible states: "Remove not the ancient landmark, which thy fathers have set." There are many other references in the Bible which refer to tools such as measuring lines and plummets, and to the units of measure that were in use at the time.

Even earlier evidence of land measurement, which we now know as surveying, occurred in the Babylonian empire about 2500 B.C. Archaeologists have found examples of maps made by early Babylonian scribes. These maps indicate that some method of land measurement and record keeping was in use. Archaeological studies of the building ruins of the Babylonian and early Hebrew civilizations indicate that some type of building layout system must have been used as well.

In early Egypt, parcels of land were measured for taxation and other purposes. There is evidence that land boundaries often had to be reconstructed using strategically placed permanent markers after the seasonal flooding of the Nile River.

Of course, the engineering and construction accomplishments of the Egyptians, Greeks, and Romans are well known. Remarkably preserved examples of crude but effective surveying and building layout tools that were used by these civilizations are preserved today in museums around the world.

In more recent times, studies of American history and the biographies of men such as George Washington, Thomas Jefferson, and Abraham Lincoln reveal that these men, as well as other famous Americans, worked as land surveyors at some point in their careers. Meriwether Lewis and William Clark of the famous Lewis and Clark expedition were commissioned by the government to explore, survey, and map the country to the west of the Missouri and Mississippi rivers. Often these early explorers and surveyors made discoveries of immeasurable importance to the development of the western United States.

However, most of these early pioneer surveyors were only interested in measuring and mapping the land. None of them were concerned with surveying as it applied to construction projects. In those times there were simply no large construction projects in progress that would warrant the development of construction surveying and building layout as we know them today.

However, this was all to change and soon. Architects and engineers began to develop design techniques and building materials which made the construction of taller buildings and longer bridges possible. To take advantage of the latest technological advances in design and materials, engineers had to develop more precise measurement and control techniques. The new construction control techniques employed the very latest surveying equipment available at the time.

When the Civil War broke out, the construction of fortifications, ports, and transportation systems became high priority projects in both the North and the South, and the use of the latest surveying techniques and their construction applications were put into immediate use.

The most extensive and the most precise use of nineteenth-century construction surveying practices occurred shortly after the Civil War during the construction of the transcontinental railroad and the Brooklyn Bridge. The Panama Canal, the Empire State Building, and the Hoover Dam projects brought the science of construction surveying and building layout into the twentieth century.

Today, the complexity of buildings, the dimensional accuracy required in their construction, zoning laws, and the cost of land have made extremely accurate building layout very important. Serious legal problems can arise if a structure is not built exactly where and how the architect, engineer, and owner want it built.

1-2   Different Types of Surveys

Throughout this text, reference will be made to various types of surveys and their purposes. The following definitions will serve to clarify future references.

Plane Survey

A plane survey is one which covers a relatively small area, no more than a few square miles at the most, in which all horizontal measuring is done in a single horizontal plane. When a change in elevation occurs, the horizontal measurement is treated as if the area were flat. For example, when measuring the linear distance from a point on a mountainside to a point in the valley, the fact that there is a change in elevation is ignored. The distance is measured in steps, with only the horizontal distances recorded. This type of surveying is the one most often used.

Geodetic Survey

The geodetic survey is similar to the plane survey, except that greater distances are covered. Like the plane survey, horizontal measuring is done in one plane, usually referenced to a certain distance above mean sea level (MSL). However, because the distances measured are so great, the curvature of the earth is taken into account mathematically and the horizontal distances and angles are adjusted accordingly. State line boundaries, international borders, and some highway and railroad surveys use geodetic techniques to ensure a high degree of accuracy.

Topographic Survey

This type of survey records the horizontal plane measurements over the surface of the earth, but it also includes indications of changes in elevation as well. A topographic map drawn from a topographic survey will have lines (called contour lines) drawn on it which represent the various elevations within the boundaries of the survey. Depending on the size of the area involved, the topographic survey may be plane or geodetic with reference to the horizontal measurements involved.

Property Survey

Property surveys, sometimes called boundary or cadastral surveys, are the oldest type of surveys in use and are probably the most common type of plane survey performed. The property survey is used to locate property lines and corners, to determine the area of parcels of land, and to plan and lay out subdivisions of property. The property survey is usually required by law when real estate transactions take place transferring ownership from one person or company to another. The property surveyor will usually produce a written legal description as well as a map or plat of the property surveyed. These descriptions and plats will often be recorded in official government records maintained in local courthouses and tax offices.

Route Survey

Route surveys are surveys along the route of highways, railroads, pipe lines, power lines, and so forth. Information gathered from a route survey will often include elevation notes, location and ownership of structures, earthwork requirements, and other information as requested by planning and engineering officials.

Hydrographic Survey

The hydrographic survey is similar to the topographic survey, but rivers, lakes, and other bodies of water are the main object of the survey. Elevation lines both above and below the surface of the water are usually shown.

Construction Survey

Construction surveys are for the purpose of planning a construction project, locating and laying out the building, and controlling the dimensions of the structure during the construction process.

As-built Survey

The as-built survey is used to accurately produce plans and maps after a project or portions of a project have been completed. Often during construction, dimensional changes are made which must be recorded and new plans must be drawn. The as-built plans may differ considerably from the original plans.

Preliminary Survey

The preliminary survey is done for planning purposes. This survey will show the location of existing structures and natural terrain features. Some preliminary surveys are drawn to a scale and to such a level of accuracy that the location of individual trees and shrubs is shown.

There are many other types of special surveys, each having its own particular use. Among them are control, city or municipal, marine, mining, forest, crop, and archaeological surveys. Each type of survey requires a different approach and a different level of skill and training.

1-3   Bearings and Azimuths

Surveyors, contractors, engineers, architects, pilots, ship captains, and many other professions frequently need to communicate compass directions to other people. The compass directions used in these communications can be used to describe the direction in which someone is traveling, the direction along a property line, or the direction down the center line of a highway or airport runway.

Almost all written and verbal methods used to describe a compass direction are based on a circle which is divided into several parts and sub-parts. Some European countries use a method in which a circle is divided into 400 parts called gons or grads. Some military units use a system which divides a circle into 6400 parts called mils.

The two systems most commonly used in the United States and much of the rest of the world are the bearing system and the azimuth system. Both methods refer to a horizontal circle inscribed on the plane of the survey or map being used. The circle is divided into 360 parts called degrees. Each degree is broken into 60 parts called minutes, and each minute is broken into 60 parts called seconds. Occasionally, the smaller divisions of the degree are noted as decimal degrees. Thus, 30 minutes 0 seconds would be the equivalent of one half of a degree and would be expressed as ".5 degrees."

The bearing method uses a circle divided into four quadrants. The quadrants each span 90 degrees and begin and end on one of the four major compass points: north, south, east, or west. Therefore, each quadrant forms a right angle and covers one quarter of the circle.

Bearings

There are six rules used in expressing direction by bearings:

1. North and south are considered to be at 0 degrees.
2. East and west are considered to be at 90 degrees.
3. A bearing is always less than 90 degrees.
4. The first direction noted in a bearing is either north or south.
5. The second direction noted in a bearing is either east or west.
6. If the direction expressed is one of the four primary directions (N, S, E, W), only the primary direction designation is customarily used.

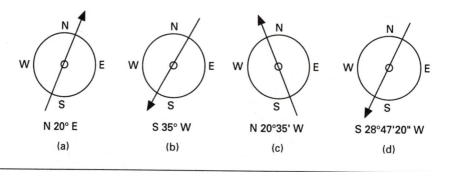

N 20° E   S 35° W   N 20°35' W   S 28°47'20" W

(a)   (b)   (c)   (d)

Figure 1-1.

Figure 1-1 shows several examples of directions expressed as bearings and the direction of traverse (travel) associated with each bearing.

In each of the examples in Figure 1-1, the first letter, N or S, indicates the primary direction. The second letter indicates the secondary direction. In example (a), the primary direction is N for north and the secondary direction is E for east. In example (a), the 20° term indicates how many degrees **from** the primary direction (north) to turn **toward** the secondary direction (east).

To see how this works in practice, face the primary direction, in this case north. Next, turn 20 degrees toward the secondary direction (east) and **stop**. You should be facing slightly to the right of north.

To practice example (b), face south, then turn 35 degrees to the west.

Examples (c) and (d) can be done the same way. The only difference is that you would need a good-quality surveying instrument to accurately measure the angles down to the smaller divisions of minutes and seconds.

Azimuths

Azimuths are a method of expressing direction using the entire 360 degrees of a circle, rather than breaking the circle into four quadrants as is done when the bearing method is used. The only thing to remember is that *azimuth angles are always measured clockwise unless specified otherwise and 0/360 is always north unless specified otherwise.*

Figure 1-2 shows examples of azimuth headings of (a) 20 degrees and (b) 234 degrees.

The direction of 20 degrees of azimuth is simply 20 degrees to the right of

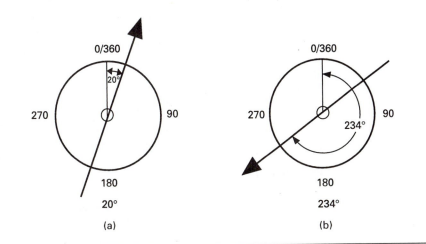

Figure 1-2.

north. To find the direction 20 degrees, simply face north, turn to the right until you are facing the 20 degree mark of a circle whose 0/360 point is oriented toward the north.

To find the azimuth direction of 234 degrees, face north, turn to the right, past south, until you reach the 234 degree mark, **or** turn to the left 126 degrees until you reach the 234 degree mark.

$$
\begin{array}{r}
360° \\
-234° \\
\hline
126°
\end{array}
$$

CONVERTING BEARINGS TO AZIMUTHS—The following chart may be used for converting bearings to azimuths.

Bearing	to	Azimuth
a. N X° E	=	X°
b. S X° E	=	180° – X°
c. S X° W	=	180° + X°
d. N X° W	=	360° – X°

EXAMPLE

See Figure 1-3(a). Find the azimuth equivalent of bearing S 35° W.
    Using formula c and substituting 35° in place of the X° value, add 180° to 35° to find the azimuth of 215°.

CONVERTING AZIMUTHS TO BEARINGS—The following chart may be used for converting azimuths to bearings.

Azimuth	to	Bearing
a. If X° = 0° to 90°	=	N X° E
b. If X° = 90° to 180°	=	S (180° – X°) E
c. If X° = 180° to 270°	=	S (X° – 180°) W
d. If X° = 270° to 360°	=	N (360° – X°) W

EXAMPLE

See Figure 1-3(b). Find the bearing equivalent of an azimuth of 263°.
    Using formula c and substituting 263° in place of the X° value, subtract 180° from 263° to find the value of 83°. The bearing would be S 83° W.

Until one is experienced in working with both bearings and azimuths and can convert from one to the other easily, it is a good practice to sketch the problem to aid in visualizing the compass directions involved. Sketch a compass face showing the primary points on a piece of scratch paper, an unfinished floor, or in the dirt if necessary. Label the primary compass points, face the primary direction of the bearing, and turn toward the secondary directions. Your position relative to the major compass points should assist you in visualizing the problem and in reaching a correct solution.

Remember that when you are adding or subtracting minutes and seconds you are dealing with a whole unit of 60 instead of 100. It works just like adding and subtracting time. Some engineering and scientific model pocket calculators have the capability of performing mathematical functions using degrees, minutes, and seconds and decimal degrees.

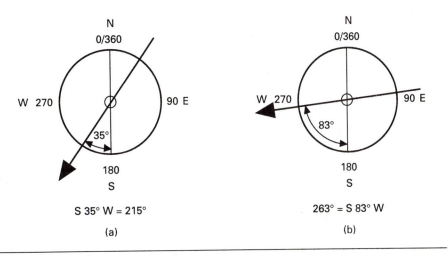

Figure 1-3.

1-4   Compasses

Explorers, sailors, soldiers, and surveyors have used compasses ever since it was discovered that a magnetic needle suspended or supported at its center would point in a northerly direction. Compasses have evolved, as have all instruments, from the first crude inventions to highly sophisticated instruments capable of extreme accuracy. Modern instruments used in navigation and surveying are capable of giving directions accurate to within fractions of a second of arc. Some modern direction- and position-finding instruments no longer use the principle of the magnetic compass. These instruments use gyroscopes and satellites to locate points on the surface of the earth and to provide accurate directions for surveyors, navigators, engineers, and scientists.

Most construction field engineering layout work is performed with reference to a fixed set of points which have been established prior to the start of the project. However, construction field engineers may occasionally have to use compasses to verify directions when laying out buildings, identifying property lines, or when doing topographic surveys. The accuracy requirements of the layout work to be performed will dictate whether a compass heading is accurate enough for the job or whether another more accurate method, such as a Polaris observation, is needed to establish direction.

Most manufacturers of surveying equipment have a variety of magnetic compasses available in several price ranges. Many of these compasses resemble those used by soldiers and Boy Scouts and are quite simple to operate and read. However, the tube compass and the surveyor's compass require some explanation.

Tube Compass

This compass is a short tube, two to three inches in length and about one inch in diameter. It is usually mounted horizontally on or in one of the uprights of a surveyor's transit. The south end of a magnetic needle suspended in the center of the tube is visible through a viewing port in one end of the tube. The south end of the needle is aligned with marks on the glass of the viewing port. This indicates that the instrument is oriented toward magnetic north. No direction readings are taken from the compass; instead, they are taken from rotating circles on the body of the instrument after the instrument has been properly oriented. See Figure 1-4.

Figure 1-4. Tube compass.

Surveyor's Compass

The main difference between a conventional compass and a surveyor's compass is in the design of the face. On the face of a surveyor's compass, the E and W are reversed from their true relative positions. See Figure 1-5. This enables the surveyor to read directions directly as a bearing.

For example, if the direction we want to find is 30° to the right of north (30 degrees of azimuth), then the bearing would be N 30° E, a north easterly direction. The surveyor's compass would be oriented so that the needle is pointing toward a point on the ring 30° to the left of north. A line extended from the south (180°) point on the compass face through the north (0–360°) point on the compass face would be the 30 degrees of azimuth or the N 30° E line. This places the north end of the needle in the quadrant between the N and the E. This quadrant is **not** the true northeast direction quadrant, but is labeled this way so that the surveyor can read the bearing N 30° E directly from the face of the compass without having to perform any calculations. This reduces the chances of the surveyor making an error in reading a bearing.

The ring on the surveyor's compass may be divided into 360° for reading direction as an azimuth or it may be divided into four quadrants of 90° each for reading direction as a bearing. Surveyor's compasses may be special ordered with two rings, one with azimuth graduations and one with bearing graduations.

Figure 1-5. Surveyor's compass. (Photo courtesy of Sokkia Corporation.)

Surveyor's compasses are usually found in the center of open frame transits, but they can be purchased separately for use as a hand-held instrument.

NOTE:

Most compasses are equipped with a mechanism that will raise the needle off its base and lock it so that it will not move. This procedure is called "slaving" or "caging" the needle. The needle is slaved or caged to protect it and its pivot while the compass is in storage or when it is being moved from one place to another. The needles and pivots on some compasses contain jewels that are subject to wear and damage if this procedure is not followed.

Because all compasses used in construction and surveying are expensive and delicate instruments, the field engineer should read and follow the manufacturer's directions when using the compass.

1-5   Magnetic Declination

While it is true that magnetic compasses will indicate a northerly direction, the direction indicated is magnetic north, not true north. The magnetic pole is not located at the true "top" of the world, so in the easternmost regions of the United States the compass needle will point to the left of true north. In the western United States, the compass needle will point to the right of true north. Figure 1-6 is an "isogonic map" that shows the approximate angles of declination for all areas of the country as they were positioned when this map was made.

Unfortunately, the angles of declination are not fixed. Their location changes with time. Also, the rate of change at any given location may vary unpredictably. This presents a problem for the surveyor if he or she is dealing with old surveys which were based on magnetic headings. There may be as much as 10 degrees or more difference in the magnetic heading of 100 years ago and the same magnetic heading today in the same location.

The field engineer must be aware of the phenomenon of magnetic declination and its deviations in order to make allowances for what might otherwise be thought to be inaccurate surveys. However, if after he or she has considered the angle of declination problem and the boundaries or other lines still appear to be incorrect, the field engineer should report these concerns to his or her supervisor. The services of a professional surveyor or engineering company may be required to check any lines that may be in question.

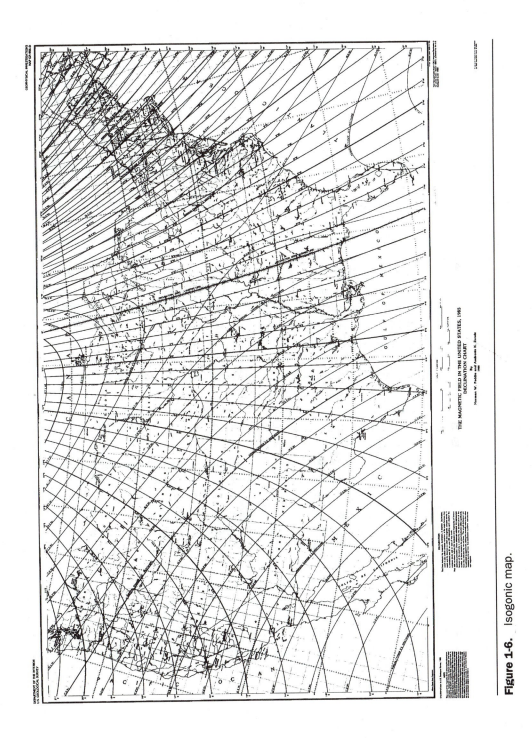

Figure 1-6. Isogonic map.

⊕ **NOTE**

Isogonic maps and other maps of interest to the surveyor and the field engineer can be ordered from the U.S. Department of the Interior, Geological Survey, Box 25286, Denver, Colorado 80225.

1-6   Metes and Bounds

Expressing distances and angles, both graphically and in writing, is the primary means of communication for the surveyor. The boundaries of a parcel of land may be shown graphically by drawing a map called a "plat" or may be described in writing by a technique known as "metes and bounds." The written description, metes and bounds, is also sometimes called the "legal description." These written descriptions are found in deeds, wills, and other legal documents in which property must be precisely described. These documents are often kept on file in courthouse records, legal and real estate offices, and other legal-document depositories across the country.

The metes and bounds are usually written by lawyers, surveyors, or others specially trained to do this work. Even the metes and bounds for a small piece of property may be quite complicated, so the people who write metes and bounds must take great care to produce error-free work. One small mistake in writing metes and bounds can result in major legal problems for the landowners for many years.

Reading and interpreting metes and bounds takes some training and practice as well. Anyone responsible for interpreting metes and bounds must realize that there are many ways to describe the boundaries of a parcel of land. The words and phrases used by different writers might sound different, but properly drawn plats using the metes and bounds produced by different writers would show identical property boundary lengths and angles. Also, a qualified person should be able to find the property and stake out its boundaries using any of several different-sounding descriptions.

An associated problem of interpreting metes and bounds is that of changes in the use of terms and phrases over time. The words and phrases used in metes and bounds written as recently as the 1800s may sound quite foreign to a young surveyor today. Fortunately, these strange-sounding phrases are easily interpreted, and one quickly becomes familiar with the terms used in a particular region.

Reproduced below is part of a legal description of a piece of real estate located in Erath (pronounced E-rath) County, Texas. The original survey was done and the description written in the late 1800s.

> Situated in Erath County, State of Texas, a part of the James Parclay League; Beginning at a stone mound in a spring branch on the North line of said Parclay survey, 986 varas from its N.W. Corner, from which a live oak tree brs. North 7 degrees west 14 varas, and another bears South 51 1/2 degrees E. 8 vrs.; Thence South 1466 vrs. to a stone mound from which a post oak bears . . .

This description is typical of early surveys. Little thought was given to the fact that the trees and stone mounds might be destroyed over the course of time by the elements. The directions were given to the nearest one-half degree in one case and the linear distances were given in varas, an old Spanish unit of measurement. The vara is explained in some detail later in this chapter.

A legal description written in the 1990s is show below.

> BEGINNING at an iron rod found for corner, said point being S. 44 deg. 24 min. 17 sec E. a distance of 1050.37 feet and S. 47 deg. 00 min. 02 sec. W. a distance of 594.93 feet from the Northmost corner of the E. Turner survey:
> THENCE S. 47 deg. 00 min. 02 sec. W along and near a fence a distance of 476.03 feet to an iron rod found for corner;
> THENCE . . .

When the metes and bounds written in the 1890s is compared with the one written in the 1990s, the difference is quite obvious. In the later description the distances are in decimal feet and the directions are shown to the nearest second. This represents the transition from the old to the new in the use of techniques and equipment. Also, the newer survey notes that an iron rod was "found for corner." This indicates that that particular point had been marked during some prior survey when iron rods were readily available. Stone mounds and trees are usually not used as markers and reference points in more modern surveys.

Some surveys are not run precisely along property lines. In cases where the terrain or other obstructions make it impossible, survey lines are offset and run parallel to the true line. The term "offset for corner" is used frequently. For example,

THENCE East 588.7 feet to corner in middle of creek, an iron stake for marker offset on westerly bank of creek; THENCE in a Northeasterly direction . . .

In the case described above, it is obvious that the surveyor could not place the corner stake in the center of the creek, so it was placed on the property line on the bank of the creek. Corner markers are often offset to the side of a road if the true line runs down the center of the road.

Some corners are not marked at all. In many cases, it is impossible or undesirable to mark corners, so other nearby points are marked or referenced and the corner is merely described by referring to other points. These points are called reference points or witness points.

In all surveys which are done to re-establish old corners or lines, the surveyor is bound by ethics and laws to do all that is possible to follow the old line. Even if the old lines are proven to be many degrees off what is now known to be a certain direction, the intent of the original surveyor is the deciding factor in most cases. Determining the intent of the original surveyor is sometimes difficult if the original lines or marks have been lost or destroyed.

Because of the legal aspects of attempting to reconstruct old surveys and the expense of new surveys, most states have laws on their books that make it a crime to knowingly disturb any survey marker. There are cases on record in which survey markers were moved by people who stood to gain if a road, dam, or other structure were built in a different location from where it was intended.

Varas

In many parts of the southern, southwestern, and western United States, an old Spanish unit of measure called a "vara" or "Spanish yard" was originally used to lay out parcels of land. There is historical evidence that as many as seven or more varas of different length were used in the territory occupied by the Spanish. Each vara was approximately 33 inches long.

To avoid disputes involving land sales among early settlers, most surveyors found it easier to lay out the parcels and write the metes and bounds using the vara length that was in use where the land was located. This practice continued in some areas well into the twentieth century, when most states passed laws that standardized the length of the vara in their area. To determine the legally recognized length of a vara in any given location, state or county surveying offices should be contacted.

Chains

Another unit of measure commonly used in early surveys, especially in surveying government lands, was the "chain." The chain was actually a heavy iron chain 66 feet long. It was made of "links," each link being 7.92 inches in length. Many older land descriptions and plats will specify distances as being so many "chains and links" in length. Often, the chain and link terms were abbreviated.

For example, in one part of the legal description of a parcel of land, the direction and distance along a part of the property line might be noted as:

..., thence N 56°20' E, 23 ch. 16.56 lks ...

This means that from a certain point, the property line would run in a direction 56 degrees plus 20 minutes to the right of north for a distance of 23 chains plus 16.56 links. The length in chains might also be shown with the links expressed as a decimal added to the chains. In this case the length would be 23.1656 chains.

To convert chains and links to feet, express the links as a decimal added to the chains, then multiply by 66.

EXAMPLE

23.1656 ch. × 66 = 1528.9296 ft

NOTE

If the measurement above had been 23 ch. 6.56 lks., the conversion would have been 23.0656 ch. because one link is 1/100 of one chain. Therefore, 6.56 lks. is 6.56/100 of one chain. The fraction 6.56/100 is converted to the decimal .0656 ch. This decimal is then added to the 23 chains. The result would be 23.0656 ch. or 1522.3296 ft.

Other Measurement Units

Historically, those who were responsible for writing the legal descriptions of property used the units of measurement which were most common in the immediate area or the unit in which the writer was trained or was the most familiar with. Units such as rods, hectares, furlongs, and so forth have all been used. Appendix A contains many of these units of measurement, along with their equivalents and formulas for conversion. Metric conversion tables are also contained in Appendix A.

The construction field engineer must often read and interpret metes and bounds. This may be necessary for site planning purposes or for locating property corner markers previously placed by surveyors but which may have been overgrown or covered up. The most common way to approach the interpretation process is to draw a map or plat of the land which is described by the metes and bounds.

A person with experience in metes and bounds interpretation can quickly read through a legal description and have a good idea of the necessary scale and starting point for the map. The novice interpreter should start in the center of a large piece of paper and use a small scale until the general size and shape of the land is determined. Then a second map may be quickly drawn using the scale most suited for the paper size and detail necessary. The starting point relative to the completed map will be readily apparent.

Of course, most computer graphic programs will accept line and angle entries from the keyboard. Certainly any good surveying program will. This makes the rendering of maps from metes and bounds much easier. Once the computer operator has entered the description in the proper sequence, the map can then be plotted to the desired scale with considerable accuracy.

Figure 1-7 shows the legal description and the resulting plat of a fictitious piece of land. As an exercise, the student is encouraged to read the legal description and follow the point-by-point development of the plat at the same time.

Note that one of the linear distances is in chains and another is in varas. Rarely, if ever, would varas, chains, and feet all be used in the same description, but in this case all three have been used for demonstration purposes. The length of the vara used in this example is 33.333 inches.

Also, note that the bearings of each line have been shown as well. Again, this is not normal procedure, but has been shown here for clarity.

If the legal description is part of a deed or other legal document, the legal description will usually be preceded by a paragraph which tells where and in what book the document is recorded. In some cases, a concluding paragraph

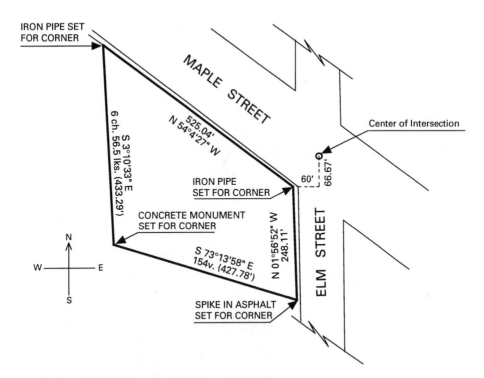

From a point at the center of the intersection of Maple and Elm streets, thence South 66.67' to a point, thence West 60' to a point, this is the notheast corner of this survey and the point of beginning, an iron pipe set for corner;
THENCE N 54°4'27" W, 525.04' to a point, an iron pipe set for corner;
THENCE S 3°10'35" E, 6 ch. 56.5 lks. to a point, a concrete monument set for corner;
THENCE S 73°13'58" E, 154 varas to a point, a spike driven in the asphalt for corner,
THENCE N 1°56'52" W, 248.11' to the point of beginning.

Figure 1-7.

will be added to the legal description which gives the amount of square feet or acres contained in the survey.

1-7   Math and Trigonometry Review

It is hoped that the reader has completed some study of mathematics, preferably courses in algebra, trigonometry, and geometry, prior to reading and using this text. However, as that may not always be the case, this section is presented as both an initial introduction as well as a refresher in the most basic mathematics required in the use of surveying techniques for construction projects.

Right Triangle

Perhaps the most-used geometric shape in design and construction is the right triangle. This is simply a triangle with one "square corner" or 90° angle. By using this basic shape and the mathematics associated with it, the designer and the builder can control many of the angles and directions necessary for the completion of a construction project.

Mathematicians label the sides and angles within triangles. This helps them in writing equations about triangles and in describing triangles to others without having to draw each triangle they describe.

Study Figure 1-8a, which shows a right triangle with its labels. Note that the upper-case letters are used to label the angles and the lower-case letters are used to label the sides. The side opposite the right angle (C) is always the hypotenuse. The other sides may be remembered as "a" for altitude and "b" for the base. Side "a" is sometimes called the "h" side (for height). Note that the letters labeling the sides and angles are opposite each other. That is, side "a" is on the opposite side of the triangle from angle A, and so forth. The right angle in the triangle is sometimes shown with a small square drawn in it. This just tells the observer that this particular angle is indeed a right or 90° angle.

When describing a triangle, mathematicians refer to the various sides and angles in different ways. For example, the horizontal base of the triangle may be called side b, side CA (because it runs between angles C and A), the side opposite angle B, or the side adjacent to angle A.

Laying out a Right Angle

Of all the right triangles with their many different possible combinations of width and height, probably the most-used right triangle is the 3-4-5 triangle.

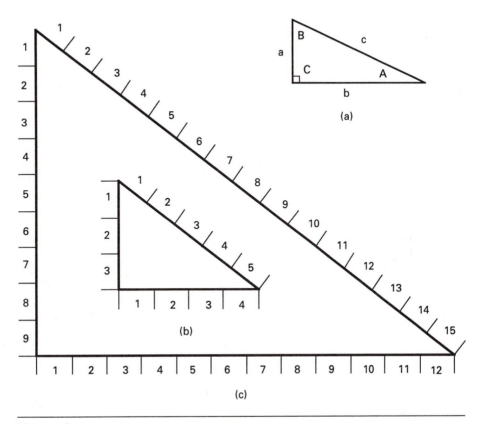

Figure 1-8.

See Figure 1-8b. The sides of this triangle are 3 units, 4 units, and 5 units in length. The units may be inches, feet, yards, meters, or any standard unit of measure. If the three sides are carefully measured and the end points of each side properly connected, there will always be one 90° angle within the triangle.

The same is true if the length of each side is multiplied by a common number. For example, multiply each side by 3. See Figure 1-8c. The sides now measure 9 units, 12 units, and 15 units in length but there would still be one 90° angle within the triangle. The 3-4-5 **ratio** still exists. Only the lengths have changed.

Using this principle and a rule or tape, the field engineer can lay out a 90° corner from any point on a line.

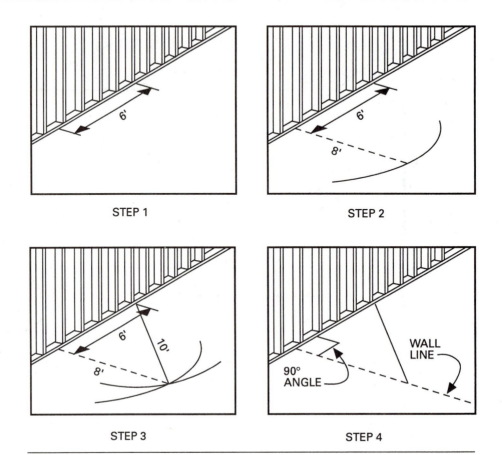

Figure 1-9.   (Illustration by Dale Barnes.)

EXAMPLE

A field engineer must establish a line along which an interior wall will be constructed. The interior wall intersects an exterior wall at a 90° angle and the intersect point is known. Each step in solving this problem is outlined below. Also, Figure 1-9 illustrates these steps.

STEP BY STEP APPROACH

Personnel required: Field engineer and one assistant.

Tools required: Measuring tape and pencil.

Step 1. From the known intersect point, measure along the base of the exterior wall a distance of 6 units. (We will use feet in this example.) Place a mark on the floor at the base of the wall.

Step 2. From the intersect point on the exterior wall, estimate a 90° angle and measure out 8 feet from the existing wall. Scribe a short arc at the end of the 8-foot line. **Note:** At this point, you are just estimating the 90° angle. Do not worry about it not being exactly 90°. Just come as close as you can.

Step 3. From the 6-foot mark you established in step 1, measure 10 feet toward the arc you scribed in step 2. Scribe another short arc at the end of the 10-foot line. The arcs should cross. Note: If the arcs do not cross, repeat step 2, drawing a longer arc.

Step 4. Draw a line from the intersect point through the point where the arcs scribed in the previous steps cross. If you have followed the proper procedure, this line will intersect the exterior wall at approximately 90°.

Step 5. Inspect your work.

A. Stand back and look carefully at the line you have established. Does it look right? Does it appear to truly intersect the wall at a right angle?

B. Double check your work with a carpenter's square.

C. Double check the prints for the location of the intersect point. Is it correct?

D. Is the line you just established the center line of the wall or is it the edge of the finished wall? If it is the center line, you should mark it as such. If it is an edge line, mark it, showing which side of the line the wall is to be built on, or lay out a parallel line corresponding to the other side of the wall.

"Pythagorean Theorem"

The field engineer will not always have the leisure of being able to perform layout work using the 3-4-5 triangle. However, the right triangle (one with one 90° angle) may still be used even if the length of **one** of the sides of the triangle is not known. In this case, the "Pythagorean theorem" may be used to determine the missing dimension, thus allowing the field engineer to continue to use the right triangle in layout work.

The formal statement of the Pythagorean theorem is something like this:

A square constructed on the hypotenuse of a right triangle is equal to the sum of squares constructed on the other two sides.

This statement may sound complicated, but it is really quite simple. Study Figure 1-10. Notice that the number of small squares in the square sitting on the hypotenuse is the same as the total number of small squares on the other sides.

The Pythagorean theorem, as written out above, is too long and compli-

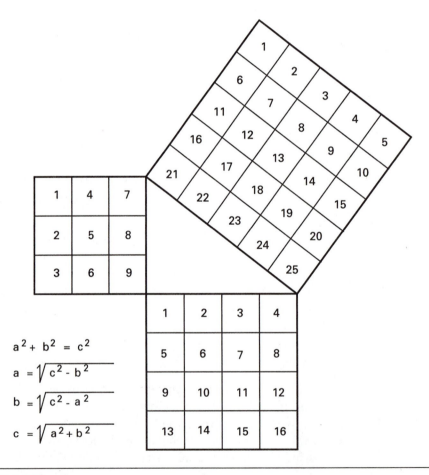

$$a^2 + b^2 = c^2$$

$$a = \sqrt{c^2 - b^2}$$

$$b = \sqrt{c^2 - a^2}$$

$$c = \sqrt{a^2 + b^2}$$

Figure 1-10.

cated for everyday use, so it is usually written in mathematical shorthand (as a formula):

$$a^2 + b^2 = c^2$$

Applying this formula to Figure 1-10, the lengths of sides a, b, and c are substituted into the formula. This gives:

$$3^2 + 4^2 = 5^2$$

This is further developed to:

$$9 + 16 = 25$$

Compare the results of this calculation to Figure 1-10.

The formula $a^2 + b^2 = c^2$ can be re-arranged and used in several different ways. Figure 1-10 shows several ways to use this formula, depending on which side of the triangle is unknown.

By selecting and using the proper formulas from Figure 1-10, the field engineer can construct a true right triangle for use in layout work.

Trigonometric Functions

A field engineer can also use the right triangle in layout work even if only the length of **one side of the triangle and one angle** (other than the right angle) within the triangle is known. This method requires an understanding of a triangle "code" called trigonometric functions. This "code" describes the mathematical relationships between the angles and the sides of a triangle.

For example, if one side of a right triangle were twice as long as another side, the relationship between the two sides would be 1 to 2. Of course, the mathematician could divide 1 by 2, getting .5, and say that the relationship between the two sides is simply .5. It makes no difference how long the sides of the triangle are; as long as one side is 2 times longer than the other, the .5 relationship will remain constant. Not only does the relationship remain constant, but **all the angles within that right triangle remain constant as well.** These fixed relationships between side length and angles are called trigonometric functions.

This fixed relationship makes it possible to determine the length of any side of a right triangle if the length of one other side and one angle (other than the right angle) is known.

The right triangle has six possible combinations of relationships among its three sides. Each relationship has its own special name—sine, cosine, tangent,

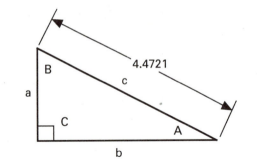

$$\text{SIN angle A} = \frac{a}{c} \qquad\qquad \text{SEC angle A} = \frac{c}{b}$$

$$\text{COS angle A} = \frac{b}{c} \qquad\qquad \text{CSC angle A} = \frac{c}{a}$$

$$\text{TAN angle A} = \frac{a}{b} \qquad\qquad \text{COT angle A} = \frac{b}{a}$$

NOTE: The same relationships exist for angle B.
Just substitute "B" into the above formulas.

Length of side a = c SIN A or b TAN A or c COS B or b COT B
Length of side b = c COS A or a COT A or c SIN B or a TAN B
Length of side c = a / SIN A or b / COS A or a / COS B or b / SIN B

Figure 1-11.

secant, cosecant, or cotangent. These are usually abbreviated as sin, cos, tan, sec, csc, and cot. Figure 1-11 shows the six combinations and the sides of the right triangle each represents.

Tables of the values of trigonometric functions for all possible combinations of side length ratios and the angles associated with them are published in trigonometry texts, surveying texts, and other mathematics references. Also, many pocket calculators have the ability to generate the numerical values of sin, cosine, and tangent ratios. A table of trigonometric functions is found in Appendix B of this text.

Here is an example of how to find the length of one side of a right triangle if only one other side and one angle is known.

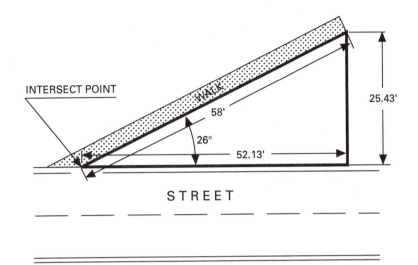

INTERSECT POINT

WALK

58'

26°

52.13'

25.43'

STREET

Figure 1-12.

EXAMPLE

A walk 3 feet wide and 58 feet long must be built. The right side of the walk intersects a street at a known point at an angle of 26 degrees to the left. The field engineer must lay out the location for the concrete forms for this project. Figure 1-12 shows the layout of this job.

STEP BY STEP APPROACH

Personnel required: Field engineer and one assistant.

Tools required: Measuring tape, stakes, hammer, piece of cardboard or heavy paper approximately 2 ft square, large nails, marker.

Step 1. Visualize the walk as the hypotenuse and the street as side "b" of a right triangle. Find the length of the "b" side. From Figure 1-11, select a formula which will give the length of side b, given the angle A and the length of the hypotenuse. c COS A will do. This means that you should multiply the length of c (58 ft) by the cosine of angle A.

Look up the value of COS 26 degrees in the table of trigonometric functions in Appendix B. The value is .89879. Multiplying 58 ft by .89879 gives 52.13 ft. This is the length of the "b" side of the triangle.

Step 2. At this point, a triangle could be laid out using the method previously described in the explanation of how to lay out a right triangle. However, due to the distances involved and possible obstructions, the best way is to find the length of side a and lay it out by using a tape. To do this, choose a formula from Figure 1-11 which will give the length of side a. c SIN A will work. Again, consult Appendix B for the SIN of 26 degrees. The value of .43837 is found. Multiplying 58 ft by .43837 gives 25.43 ft.

 NOTE

Always select the formulas which use the same known values if possible. The formulas selected above both used the c dimension as a multiplier. A formula could have been selected which would have used the b dimension as a multiplier, but the b dimension was an unknown in the original problem. This practice increases accuracy and decreases the chance for error.

Step 3. From the intersect point, measure along the street a distance of 52.13 ft. Mark this point. Measure outward from this point 25.43 ft. This line should be as close to 90 degrees to the street line as you can estimate. Put a piece of cardboard or heavy paper on the ground and spike it down with large nails so that the center of the paper is at the end of the 25.43-foot line. Scribe an arc on the paper.

Step 4. Measure from the intersect point 58 ft to a point on the arc scribed in step 3. Mark this point. This is the end of the walk. Many methods may be used to mark this end point. A stake may be driven, batter boards may be erected, and so forth. These techniques are covered in later chapters. Also, methods of laying out the parallel or offset line that is the other side of the walk are covered later.

Step 5. Inspect your work. Stand back and look at the location of the lines and marks. Does it look right? Review your calculations for errors. Check the right angle with a carpenter's square or by using the 3-4-5 method.

The preceding discussion of the properties of the right triangle is intended to prepare and/or refresh the student prior to continuing further into this text. This discussion is also intended to act as a reminder and guide for future field or office activities.

However, these few pages should by no means be considered to be a complete presentation of the subject of trigonometry. There are many additional methods by which trigonometry may be used by field engineers and others involved with construction projects. The use of oblique triangles (those without a 90 degree angle) and the use of additional formulas and applicable mathematical laws are much too extensive to cover in this text. The student is encouraged to consult additional trigonometry, algebra, and surveying texts to expand his or her knowledge beyond this limited introduction.

1-8   Record Keeping

Surveyors keep accurate notes on the work they do in the field. These handwritten notes, called "field notes," are entered into small bound books as the surveyor proceeds with his or her work. A specific format has been used to record field notes for many years. The same format is used across the United States and in many other areas of the world.

After the field work is done, the field notes are used in the office to produce legal descriptions and maps of the area surveyed. The field notes are then filed and may be kept for many years for reference purposes.

When disputes arise concerning the location of property boundaries or corners, the surveyor is the "eyewitness" as to the location of the boundaries in question simply because he or she was the only one there when the point or line was established. Of course, the surveyor cannot be expected to remember the exact angles and distances of each piece of property surveyed, so all surveyors rely on field notes. During legal proceedings, the surveyor will use the original handwritten notes as references because there is a chance that a mistake might have been made when the notes were copied in the office or when maps were drawn from the notes.

If the surveyor is called to testify in court or give depositions concerning property boundaries, the field notes must be above question or they will not be credible. In many cases, field notes have proven to be the deciding factor in law suits which were filed many years after the death of the original surveyor.

Today, much surveying work is done using electronic instruments and no handwritten field notes are produced. Even with the field notes being recorded on various computer disks, tapes, or cards, courts still rely on the original records of work done in the field, except that now these records are in various types of electronic formats.

Courts try to standardize their actions in legal matters. Once a standard has been established, it is called a "precedent." Relying on handwritten documents such as surveyor's notes is such a precedent. This precedent has carried over into cases in which other types of handwritten notes have been accepted as factual legal documents. In many instances, construction project manager's diaries, journals, and daily reports have been the deciding factor in major cases.

The same is true of field engineers' recordings of layout work done on the job site. For this reason, field engineers should record their field work using techniques similar to those used by surveyors.

Field Book

The field book most frequently used by surveyors is a small bound book which will fit into a person's pocket. See Figure 1-13. These field books work quite well for field engineers, and their use is recommended. (Occasionally, loose leaf notebooks are used, but precautions must be taken to preserve the order of the pages in these books.) The field books may have various formats, but those most widely used have approximately 28 rows divided into six columns on the

Figure 1-13. Surveyor's field book. (Photo courtesy of Sokkia Corporation.)

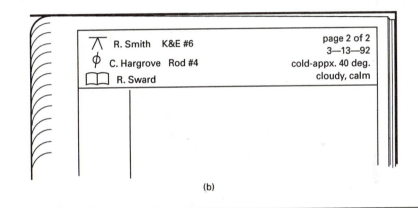

Stadia Survey — Reynolds property					
Lots 5, 6, 7 — Industrial Park					Page 1 of 2
⋏ @ B.M. EL = 100.0' H.I. = 4.982'					
BS to 1	Hor ∠	S.I.	Vert ∠	Hor. Dist.	Diff. Elev.

(a)

⋏ R. Smith   K&E #6   page 2 of 2
φ C. Hargrove Rod #4   3—13—92
▭ R. Sward   cold-appx. 40 deg.
                                         cloudy, calm

(b)

Figure 1-14.

left-hand pages and small squares (like those on graph paper) on the right. Conventionally, field data is entered on the left page and a sketch of the area surveyed is entered on the right page. See Figure 1-14.

Below is a list of the recommended procedures for recording field notes.

1. The name, address, and phone number of the person using the field book should be recorded in ink on the inside and outside front cover of the book. Some surveyors include an index to the book in this area.
2. All data entries and sketches should be in 2H or 4H pencil. Erasures are not permitted. Mistakes should be lined through one time and initialed.

Right-hand page entries:

3. On the top lines, enter the name of the instrument operator, rod holder, note keeper, party chief, or others involved in the survey at the top of the right-hand page. Small symbols may be used to indicate the duties of persons named. For example, a small drawing of a tripod by the name of the instrument operator may be used.

4. Also, on the top lines, enter the name and number of each instrument, rod, tape, or other equipment used. This allows the instrument to be identified and checked if errors are found which can only be attributed to instrument error.

5. Enter a short description of the weather. Approximate temperature, wind conditions, moisture, and ground conditions (wet, muddy, snow, etc.) should be noted.

6. Sketches of the area surveyed should be clear and not cluttered with extraneous detail. All sketches should be done with a template or small rule. Printed entries are recommended. The sketch does not have to be to scale, but an approximation of the size of the area covered is recommended.

7. An arrow indicating north should be drawn on the sketch. The northerly direction should be to the top of the page.

8. The sketch may be entered with the spine of the book up and the bottom of the book to the left if necessary.

Left-hand page entries:

9. Enter the information which will identify the survey at the top of the left page. Information included here would be items such as property identification, type of survey, and the date.

10. Label the columns to identify the data to be entered. Other columns may be drawn using the rule.

11. Enter the field data as it is being gathered. The recorder should repeat the data as it is spoken by whoever is taking the readings. The person reading the tape or instrument should then verify that the recorder heard the data properly.

Some recorders will have all the other people in the crew initial the field book alongside their name after the data entry is complete. This is not to certify that the data is correct, but that the person was actually present and witnessed the work being done.

The pages of the field book should be identified as page one of two, page two of two, and so forth. If other pages are referenced, their location should be noted. If the notes are copied, the word "COPY" should be written across the entire page. Pages should not be torn from a field book for any reason. Unwanted pages should have a line drawn across them and the word "VOID" written across the page. If additional entries to a field book are made in the office after the field work is completed, they should be made in red ink to distinguish these entries from field-entered data.

Examples of field book data entries are shown in several places in succeeding chapters of this text. The entries applicable to each different type of survey are shown as the subject is covered.

1-9   The Field Engineer's Tool Box

A true professional is judged, among other ways, by his tools and how he uses and cares for them. Field engineers' tools range from a simple #2 pencil to an electronic instrument costing several thousand dollars. In this section, the small tools and equipment indispensable to the field engineer's work will be discussed. The instruments, rods, tapes, and other precision instruments are discussed in succeeding chapters. Personal wear and protective gear are covered in the section on safety.

Tool Boxes, Bags, and Belts

Many field engineers prefer to use a regular carpenter's tool box and belt. This proves to be quite satisfactory in most cases. The tool box is used to store and carry all the tools necessary to do almost any job. It can be locked for security and a chain can be run through the handles to lock the entire box up if necessary. These boxes are approximately 30 inches long, 8 inches wide, and 9 inches high. A carpenter's tool box usually has a slit in the top so that a carpenter's square can be carried. (The short leg of the square is allowed to protrude upward through the slit.)

The carpenter's tool belt works quite well for the field engineer. The belt has two large pouches on either side and a tape-carrying compartment in the center. It also has rings and loops to hold hammers or other tools. Many field engineers use a carpenter's belt fitted with suspenders, because the tapes and hammers used by the field engineer are too heavy to hang on a belt alone. The belt can be stored in the tool box when not in use.

⊕ **NOTE**

Carpenters' belts are available in leather or vinyl. The leather belts are more expensive, but they last longer and are more comfortable, especially in hot weather. The vinyl belts and their vinyl bags do not "breathe" and allow perspiration to evaporate as do the leather ones. This can cause heat rash and blistering of the legs and hips in the areas where the bags are worn.

The canvas bags used by masons to carry their tools are often used by field engineers to carry stakes and hammers. These work quite well and may also be rolled up and stored in the tool box when not in use. Some of these bags have shoulder straps which are useful when the field engineer must have his or her hands free to manage tapes, hammers, stakes, and other equipment.

Hammers, Hatchets, and Machetes

The field engineer is usually faced with having to drive stakes into the ground. The best tool for this is a 2- or 3-pound "sledge" type hammer with a 12- to 18-inch handle.

Some field engineers carry a hatchet instead of a hammer. A hatchet can be used to remove smaller limbs and clear light brush when necessary. Hatchets can also be used for sharpening and driving stakes. Hatchets are not as good as a 3-pound hammer for driving stakes, and care must be taken to avoid the cutting edge.

Machetes are usually found in a field engineer's tool box and are used on occasion to clear brush and other vegetation from the line of sight when laying out buildings. Machetes are available in various lengths from about 18 inches to 24 inches. When buying a machete, get one that is fairly stiff and made of high-quality steel. There are some cheap machetes available, but they will not hold an edge and they bend quite easily. Remember, a dull cutting tool is dangerous. It will not "bite" into the wood and it may glance off and hurt you or others. Also, make sure that any cutting tool has a sheath and that you keep the tool in it when not in use.

Wrenches, Pliers, Knife, and Screw Drivers

The field engineer will occasionally need some hand tools to repair and maintain equipment. A 10-inch adjustable wrench and a pair of common pliers are minimum requirements.

A knife is indispensable for cutting string lines and other duties. A medium-size good-quality pocket knife or a utility knife with a retractable and replaceable blade work well. Some prefer to use a small (3-inch blade) hunting knife which is carried in a sheath on the tool belt. The sheath knife is left on the belt and is therefore always available. It is also easier to get to and use than a pocket knife.

One medium-size cross point (Phillips) screwdriver and one medium-size common screwdriver should be included in the tool selection.

Stakes, Ribbons, Flags, Tacks, and Nails

Three types of stakes are usually used in field engineering work—hubs, markers, and guards. Hub stakes are driven into the ground until the top of the stake is at a desired elevation. Markers are used to mark line or grade, and have the appropriate information printed on the side or top of the stake. Guard stakes are tall, skinny stakes placed to the side of a hub or marker stake. Their only function is to indicate the presence of a hub or marker stake.

Hubs and markers may be purchased from building-supply sources or they can be made from readily available construction lumber. If hubs are made, they are usually made by ripping a standard pine or fir 2 × 4 length-wise. Knotty portions should be discarded. This produces a stake approximately 1½" square. The length of the stake depends on the hardness of the ground it is to be driven into, but 8 to 10 inches is the usual length. The stakes are sharpened by using a power saw to form a point. In a pinch they may be sharpened with a hatchet or machete. However, a uniform point on a stake makes it easier to drive straight. Guard stakes, sometimes called "laths," are usually purchased in bundles or they can be cut out of ¼" × 1½" material. Guard stakes are usually 24 to 36 inches long. Occasionally, it may be desirable to cut guard stakes by ripping 1½" material.

In many cases, a hardened steel bar approximately ¾" by 1¼" by 18" long is used to open a pilot hole for stakes. This bar, called a "gad," is used in hard or rocky ground when a wooden stake would be destroyed by attempts to drive it into the ground. Gads are also used when stakes must be driven into compacted fill material or through asphalt parking lots.

The ribbon material used by surveyors and field engineers is made of colored plastic. It is approximately 1¼" wide and usually comes in 300-foot rolls. Many colors are available. The field engineer should have ribbon on hand in all available colors. Different items to be laid out are often indicated by using different-colored ribbon. See Section 6-2 of this text for recommended ribbon colors for different areas of the job site.

Flags are available in the same colors as the ribbon. The flags are small pieces of plastic material, approximately 2" × 2" square fastened to stiff wire staffs. These are usually sold in bundles of 100.

Brushes are bundles of stiff plastic strands approximately 5" long. These small bundles of plastic strands are designed to be fastened to the top of hub stakes. This allows equipment blades to pass over close to the top of the stakes. The brush strands bend but spring back upright. These brushes are available in many colors.

Stake tacks are available from surveyors' supply sources. These tacks are ¾" long and have an indention in the center of the top to aid in centering and in using plumb bobs.

NOTE

It is recommended that the field engineer get a rubber or styrofoam ball and stick the tacks into it. This "tack ball" can be hung from a cord on the tool belt for easy access.

Several large nails should also be carried. These come in handy for marking elevations on wood members and temporary locations in hard ground, and so forth.

Markers, Pencils, Paint, Chalk Lines, and Scribes

Along with the usual pencils and pens necessary for daily operations in any business, the field engineer should have wide- and narrow-point felt markers available in both black and red. These should be the permanent water-resistant ink variety.

Lumber crayons or keel are used frequently for marking on pavement or concrete surfaces. These are available in several colors.

The field engineer should have cans of spray paint available in several colors. Fluorescent colors are available. Surveyors' supply companies stock these spray cans, which conveniently spray upside down.

The field engineer should have two chalk lines in his tool box. One should have red chalk in it, the other, blue. Spare chalk should always be carried. The red chalk is quite permanent and is hard to get off concrete, wood, or clothing. The blue chalk comes off fairly easily. The color used should be chosen with this in mind.

A sharp-pointed scratch awl or other sharp-pointed scribe will be needed as well. These scribes are used to scratch fine lines in painted surfaces, steel, or concrete members.

Squares, Levels, and Rules

The field engineer should have two squares in his tool box, a carpenter's square and a small try square with a 6- or 8-inch blade.

Three kinds of levels will be needed frequently—a 2-foot carpenter's level, a string level, and a hand level. (See Section 2-3 for information on the use of the hand level.)

Other than precision tapes used in layout and precision taping, the field engineer should have a 25-foot carpenter's tape on hand. A 6-foot folding carpenter's rule and a 6-inch pocket rule will also be used frequently. Some field engineers also carry a 6-foot folding engineer's rule as well. (The engineer's rule is divided into feet, tenths, and hundredths of feet. The carpenter's rule is divided into feet, inches, and fractions of inches.)

Plumb Bob, String, and Magnifying Glass

Even though most instrument cases have a plumb bob in them, a spare is useful in many cases. There are times when it is desirable to leave a plumb bob in a location for a time while other work is done elsewhere on the job site.

String is indispensable for layout work, especially when setting batter boards for concrete forms. A high-quality nylon string is recommended because it is strong and resists moisture well. This string is available in various lengths in hardware or sporting goods stores.

A ten-power magnifying glass is needed to read the vernier scale on some instruments. The magnifiers which are usually in the instrument cases are easily lost. These are relatively inexpensive and a spare in the field engineer's tool box can sometimes save the day.

Miscellaneous Tool Box Items

Many field engineers carry a variety of items in their tool boxes which they have found to be of personal value over the years. Some of these are insect repellent, clean rags, gloves, a tape grip, a small first aid kit, a roll of duct tape, plastic bags (for covering instruments if it begins to rain), notepaper or notebook, clip board, and some loose change.

Although the field engineer will occasionally need to use power tools, the proper selection and use of power tools is beyond the scope of this text. It is recommended that the field engineer seek proper instruction and become familiar with any power tool prior to attempting to use it.

In addition to the hand tools and power tools mentioned above, the field engineer will occasionally need to use accessory equipment such as barriers and signs designed to protect the working group from traffic or other hazards. The company safety representative should be contacted with regard to the use of this equipment.

1-10   Professional Ethics

A "professional" is thought of as a person who is a specialist working at a particular activity, usually for profit, while requiring little or no direct supervision. This situation requires a person who has considerable self-discipline and one who is capable of being self-directing. This freedom of choice as to work habits and methods is sometimes abused, subjecting the so-called professional to adverse publicity and, in some cases, legal action.

Almost every profession has its own professional organization which attempts to represent and/or regulate the profession to some degree. Most of these professional organizations have adopted a set of rules or "codes of ethics" which outline the desired behavior of its members, both in business and in their personal lives. Additionally, most states and the federal government have passed numerous laws in an attempt to ensure that professionals operate within certain guidelines. These codes of professional ethics and the laws passed by the various legislatures are all designed to protect the general public from being cheated by unscrupulous persons hired to perform professional services. Penalties for violation of codes and laws may range from the revocation of a license to do business in a given area to fines and imprisonment.

However, there is another cost of unethical or criminal behavior which may not be so readily apparent as the fine or jail sentence. This is the cost associated with the loss of one's good reputation. There is no way to accurately measure the direct costs of the business lost just because no one trusts or wants to be around someone who operates in an unethical manner. The phone just does not ring and the lucrative contracts just do not come along. Associates do not return calls and clients refuse dinner invitations for no reason. Ethical competitors prosper while the unethical operator goes broke and wonders why.

Other costs associated with unethical operation may be incurred. These are the unnecessary legal expenses that result when unethical operators are taken to court. Legal fees, fines, and judgments can run into hundreds of thousand of dollars.

Once a professional person has achieved a good reputation, the chances of being taken to court are greatly reduced. The plaintiffs will recognize that it is hard to convince a judge or jury that a professional person with a good reputation has acted maliciously or irresponsibly. In most cases, the plaintiff's attorneys will convince the plaintiff to give up without a fight. Often, the plaintiff will never even file a suit, recognizing the hopelessness of going up against someone with a good reputation.

Most religions have sets of mores or beliefs which encourage their members to behave ethically, much in the same way as do the professional organizations. The penalties for violating these beliefs vary but may range from the extremes of death and dismemberment to being asked by officials within the religious group to refrain from specific acts and repeat a specific prayer or chant a certain number of times.

Regardless of the legal and religious ramifications which usually result from unethical behavior, doing business ethically is simply easier and more cost effective. It is easier because there is no need to worry about having to keep track of which lie was told to whom or deciding which promise to break on any particular day. Work can go on with the whole mind being directed toward the business at hand. When ethical business and professional practices are employed, a higher-quality product is produced more quickly, more money is made in a shorter period of time, and no unnecessary legal costs are incurred.

1-11   Safety

The safety issue is applicable to all persons both on or around the construction site. The field engineer is no exception and must adhere to all pertinent safety rules, just the same as other workers. The field engineering staff must be included in all job-site safety meetings and must meet or exceed employer regulations concerning drug and alcohol use. The field engineer often works alone or with only one or two assistants. The field engineer must be capable of working unsupervised for long periods of time, so he or she is essentially in charge of his or her own job-site safety.

As previously stated, the field engineer is subject to the same hazards as other construction workers and must take the same precautions. However, there are some hazards that are particularly applicable to field engineering operations. The following is a partial list of general safety rules that apply to the field engineer. The field engineer should consult with the company safety representative for additional information concerning specific job-site hazards.

1. Dress properly. Wear clothing appropriate for the season. Avoid clothing that is too tight or too loose. Bright-colored safety vests should be worn if vehicle or equipment traffic is a hazard. If working in areas

where insects or snakes may be a hazard , high boots and other specialized clothing must be worn. Insect repellent may be necessary to repel ticks and mosquitoes. A hard hat, safety glasses, and safety shoes should always be worn on a construction site.

2. Be aware. Know what is going on on the job site. Ask the project manager if exceptionally hazardous operations are under way, such as blasting or electrical-system testing. Ask if hazardous chemicals are being used anywhere on the site. Locate alternate escape routes from the building and job site.

3. Know where to get help. Find out where the first-aid equipment is located. Locate fire extinguishers and telephones. Remember where these items are relative to your position on the job site. If using a radio for communications, find out if there is a special frequency or channel to use in case of an emergency.

4. Avoid seasonal exposure. Sunstroke and heat exhaustion are hazards in summer, as are frostbite and hypothermia in winter. Consult with your company safety representative to learn the symptoms of these seasonal hazards and their first-aid treatment.

5. Be extremely cautious around construction equipment. Make sure the operator knows you are in the area if you must go near heavy equipment that is in operation. Do not walk under loads suspended from cranes. Be extremely cautious when approaching articulated equipment such as front-end loaders. You could be caught in the center hinge mechanism. Stay a safe distance from the rear of construction equipment. The operator may reverse too quickly for you to react to back-up alarms and get clear.

6. Cautiously approach persons working with power tools. You might break their concentration, startle them, or otherwise make them lose control of the tool, resulting in injury to themselves and others.

7. Avoid open areas when thunderstorms are approaching or during thunderstorms. Many surveyors and field engineers are struck by lightning every year.

8. Do not look directly at the sun through instrument scopes unless special sun filters are used. Serious eye damage can result from momentary exposure.

9. Be careful when working around overhead electrical wires or extension cords. Do not allow even wooden surveying rods to come in contact with overhead electrical wires. Steel tapes may cut into the insulation when they are dragged across extension cords, causing a shock to the handler and damage to the equipment.

10. Be aware of open shafts and open sides of multi-story buildings. Open shafts and other elevated areas should have barriers installed, but the barriers may have been knocked down or removed for some reason. It is easy to take one too many steps backward while watching the numbers on a tape and step into an open elevator shaft.

11. Coordinate communications. Make sure all members of the field engineering crew know the hand signals or radio frequencies or channels being used for communications. A field engineering assistant may respond incorrectly to a signal or instruction from the instrument operator and, without thinking, step into the path of oncoming traffic.

12. Beware of confined areas. Do not go into a closed area for any reason unless it has been checked by the proper authorities. Ditches, tunnels, storm sewers, or other confined or low-lying and closed areas may contain hazardous vapors.

13. Do not venture into ditches or other excavations that have not been properly shored or otherwise protected from cave-in.

14. Field engineers often must go into uncleared areas. Be alert for snakes in parts of the country where they are known to live.

15. Get first-aid training. Your company safety representative or the local Red Cross can assist in arranging courses in basic first aid and cardiopulmonary resuscitation (CPR).

16. Remember, most companies today take the position that if you use drugs, you do not work for them.

17. Keep alert. Good general health is essential to all construction employees, so see your physician annually for a checkup. Get plenty of rest, and do not try to work while taking medicines that cause drowsiness. Be extra cautious in all operations if you must work overtime or irregular hours and are not as alert as you might ordinarily be. Budget your time. Work steadily but do not rush to complete a complicated job. Haste and fatigue cause mistakes that can be both dangerous and costly.

Chapter 1 Review Questions

1. Name several prominent early Americans who either once worked as surveyors or used surveying in exploring and developing new lands.

2. List three different types of surveys. What is each type of survey listed designed to accomplish?

3. Convert these bearings to azimuths:

   (a) N 36° W   (b) S 28° E   (c) S 73° W   (d) N 64° W

4. Draw a compass face with N, S, E, and W indicated. Draw arrows across the compass face indicating the direction of travel for each of the bearings in question 3.

5. Convert these azimuths to bearings:

   (a) 33°   (b) 289°   (c) 74°   (d) 195°

6. Draw a compass face with N, S, E, and W indicated. Draw arrows across the compass face indicating the direction of travel for each of the azimuths in question 5.

7. Why are the E and W reversed from their true positions on a surveyor's compass?

8. What is magnetic declination? How does it affect surveying and construction layout work?

9. What are "metes and bounds"?

10. What is a "vara"?

11. What is a "chain"?

12. Convert the following "chains" to feet and inches.

    (a) 13 ch.   (b) 8 ch. 23 lks.   (c) 5 ch. 3.5 lks.

13. Draw a 3-4-5 triangle. Properly label each side and each angle using upper-case and lower-case letters. Indicate the 90° angle.

14. A right triangle has a height of 4.93 units and a base of 8.70 units. Using the $a^2 + b^2 = c^2$ formula, calculate the length of the hypotenuse of this triangle.

15. A right triangle has a base length of 65 feet. The angle between the base and the hypotenuse is 13 degrees. What is the length of the "a" side of this triangle? What is the length of the hypotenuse?

16. Entries in a surveyor's field book are made according to certain rules and recommended practices. Why?

17. List three common ethical business practices which should be observed by all professional personnel.

18. List five common safety precautions which should be taken by all persons working on construction job sites.

C H A P T E R

2

LEVELING

2-1 Principles of Leveling

Surveys are usually done in only the horizontal plane and are keyed to a reference elevation. This reference elevation is sometimes called the "datum elevation" or "datum plane." Points on the survey are noted as being either at or a certain distance above or below the datum plane.

The most common datum plane used is "mean sea level" or "MSL." Mean sea level is determined by measuring and recording the elevation of the surface of the oceans, gulfs, and seas over a period of several years. After factoring in corrections for geographic differences, the recorded elevations are averaged. The resulting "mean" is given the elevation of 0'0". All elevations above or below this mean are said to be "above MSL" or "below MSL," respectively.

> **EXAMPLE**
>
> The highest point of land in the world is the summit of Mount Everest, where the elevation is approximately 29,028 feet above MSL. The lowest elevation in the United States, 280 feet below MSL, is in Death Valley in the Mojave Desert of southern California.

Federal and state government agencies have determined the elevations of various inland points around the country in relation to the elevation of 0'0" MSL. These points, called "benchmarks," are used by surveyors and engineers as reference elevations in planning major construction and conservation projects. Maps showing the location of these benchmarks are published by the National Geodetic Survey. These maps may be found in certain libraries that have been selected as government document depositories. Maps are also available for purchase through the U.S. Government Printing Office.

The following is an example of how a government benchmark might be used to control elevations on a construction project.

EXAMPLE

The elevation of the top of a dam for a water conservation project being planned in Arkansas is specified to be 630 ft above MSL. The nearest government benchmark is seven miles away at an elevation of 529 ft above MSL. Starting at the government benchmark and working toward the site of the proposed project, engineers would carefully measure changes in elevation at convenient points along the way.

When the dam site is reached, a job-site benchmark would be placed at a carefully selected point near the dam site. The job-site benchmark would reflect the changes in elevation between the government benchmark (529 ft above MSL) and the elevation at the location of the job-site benchmark. The contractor would then plan his work relying on the job-site benchmark elevation.

For example, if the job-site benchmark had been placed at an elevation of 600 ft above MSL, the contractor would plan his work so that the top of the completed dam would be at an elevation 30 ft above the job-site benchmark. This would satisfy the original requirements for the dam top elevation to be at 630 ft above MSL.

NOTE

The United States Coast and Geodetic Survey (now the National Geodetic Survey) markers usually show elevation only to the nearest foot. To obtain more precise information about these benchmarks, contact the Department of Commerce National Geodetic Information Center in Silver Spring, Maryland. You must provide precise information about the marker, such as the number, agency responsible, general location, and so forth. The latitude and longitude of the marker location will be helpful as well.

Some cities establish "city datum elevations." The city datum elevations are usually based on government benchmarks. The city will then place addi-

tional benchmarks in convenient locations around town. Engineers and architects use these benchmarks to plan projects throughout the city. This is of
particular importance when planning water and sewer systems so that flow
direction and flow rates can be controlled. Blueprints of proposed projects will
often have notes referring to elevations on the job site as being a certain number of feet above or below "City Datum Elevation."

On most small construction projects a government benchmark is **not** used as
a reference elevation. The architect or engineer simply picks a visible and accessible nearby point which is likely to remain undisturbed for the duration of the
project. Such an elevation point might be a spike in a utility pole across the alley
from the job site or the top of the fire hydrant on the corner of a nearby street.
This point becomes the job-site benchmark and will be given the elevation of 0.0
ft, 100.0 ft, or some other arbitrary figure. Often the job-site benchmark will be
designated as being at an elevation which will make the elevation of the finished
first floor above ground level come out to 0.0 ft or 100.0 ft. This aids in the
construction of the remainder of the building because subsequent elevation measurements can be referenced to the first-floor elevation. In this case, the job-site
benchmark may no longer be needed after the first floor is in place.

Proper leveling is also important in site grading and landscaping. Flooding
will often result if leveling is not properly done to control the flow of storm
run-off water.

Many subcontractors and suppliers fabricate building components in locations many miles from the job site and haul them to the job site for installation. If the elevations have not been carefully established through proper
leveling techniques, the various components will not fit. This results in lost
time and money.

On most construction projects, many parts of the structure are dependent
on other parts to operate or fit properly. The utility and drainage systems on the
job site must conform to the elevations of the surrounding area and meet local
community specifications concerning their operation. There may be building-
height restrictions in effect in certain areas. These items and many others are
controlled by the process of leveling. Therefore, it is of the utmost importance to
use proper leveling techniques on all phases of construction projects.

2-2   Planning Leveling Operations

Before any leveling operations can begin, the field engineer must determine just
what is required in a particular situation. Is the purpose of this leveling assignment to give grade for setting concrete forms or laying pipe? What distances are

involved? Which benchmark is to be used? These and other questions must be fully answered and understood before proceeding with the field operations.

Here is a checklist of items that the field engineer should consider before beginning leveling operations.

1. What is the purpose of this leveling operation? The purpose will often have a bearing on the level of accuracy required and the layout procedure.
2. What level of accuracy is required? It is recommended that the field engineer check with the project manager or consult the specifications to determine accuracy requirements.
3. What type of equipment is required? Accuracy requirements will often dictate the type of equipment necessary to do the job.
4. Is there a specific time during which this operation must be done? Often other activities on the job site conflict with leveling.
5. What are the site conditions? Does clearing or preliminary dirt work need to be done prior to this operation?
6. Are the prints available? What records of this operation need to be generated?
7. Is there a special color code or other special markings required on this project?
8. Who should be informed of the results of the leveling operation?

2-3   Equipment Selection

Selecting the right tool for the job has long been the mark of a true professional. This applies to the work of the field engineer just as it does to any other skilled trade. Most construction companies of any size will have several types of surveying instruments and other equipment available for use by its field engineers. It is recognized that the ideal piece of equipment for a particular job may not be available precisely when the field engineer needs it. In this case, it is better to go ahead and use a more-advanced piece of equipment that is immediately available rather than delay the job a day or two waiting for a simpler piece of equipment to become available.

All the problems of scheduling the availability of surveying equipment cannot be foreseen. Therefore, this text will take the approach that the field engineer has an almost unlimited selection of surveying equipment from which to choose, but that the equipment with the minimum capabilities necessary to do the job will be chosen.

Selecting the Instrument

Choosing the right instrument for a leveling job depends on how accurate the leveling job needs to be and how far the level reference plane must be extended. Establishing a level reference plane can be accomplished using no more than a simple pan of water and some string. However, the degree of accuracy and the distance the plane could be extended would be greatly increased if a more-advanced instrument were used.

A good general rule for the field engineer to follow in selecting surveying equipment for a particular application is to choose the least-expensive equipment capable of doing the job. This prevents undue wear and reduces the chance of accidental damage to or theft of the more-expensive equipment. Also, the less-expensive equipment is not as complicated to operate, therefore reducing the chance of the field engineer making an error.

Construction Levels

Construction levels are sometimes called "dumpy" or "wye" levels. This is a reference to their shape and the method in which the telescope is mounted on the body of the instrument. See Figure 2-1.

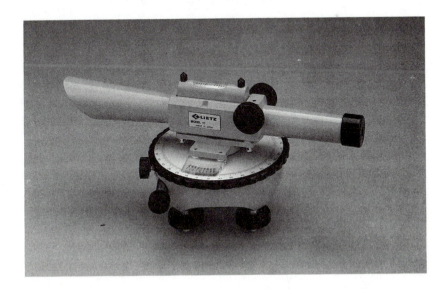

Figure 2-1.   "Dumpy level." (Photo courtesy of Sokkia Corporation.)

The telescopes used on construction levels are usually between 10X and 20X or ten and twenty "power." The X, or power, of a telescope refers to the degree of magnification produced by the lenses. For example, an object viewed through a 10X (ten-power) telescope would appear ten times larger than when viewed with the unaided eye.

Surveying instrument telescopes have a crosshair assembly mounted inside the sight tube in the lens area. The crosshairs in most telescopes manufactured today consist not of hair, but are actually fine wires mounted in a framework or attached to mounting points inside the sight tube. Some telescopes do not use crossed wires but instead use a clear glass lens which has lines etched on its surface that give the appearance of crossed wires or hairs.

The telescope lenses are made of glass or plastic and are ground to a convex or concave shape. The quality of the view through a telescope depends on the quality of the glass or plastic used and how precisely the lenses are ground. Naturally, the higher the quality of the materials and the manufacturing processes used, the more expensive the instrument.

Most telescopes will have two adjustments by which the crosshairs and the object being sighted can be brought into sharp focus. Look at the telescope in Figure 2-2. The object focus adjustment knob is usually found on the side or the top of the sight tube. The telescope is pointed in the general direction of the object to be sighted and the focusing knob is turned until the view through the

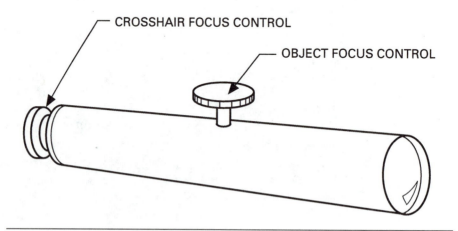

CROSSHAIR FOCUS CONTROL

OBJECT FOCUS CONTROL

Figure 2-2.   (Illustration by Dale Barnes.)

telescope is clear. Because the field of view through the telescope is relatively small, it may be difficult to "find" the targeted object by looking through the telescope. To assist in this situation, some telescopes have an open gunsight arrangement on the top or side of the telescope. By sighting across this open sight, the tube may be lined up on the target object prior to sighting through the telescope.

After the object is clearly focused, the crosshairs are focused by turning the knurled knob of the eyepiece until the crosshairs are clearly visible when looking through the telescope. At this point, the target may have become slightly out of focus, so some minor adjustment of the object focus knob may be required. Two or three successive minor adjustments of both focusing knobs may be required in order to get maximum clarity of both the target and the crosshairs.

The manufacturer's manual should be consulted for additional instructions as to the proper method of focusing the telescope of a specific instrument.

Construction levels and other surveying instruments are set to a level position by the use of bubble levels mounted on the telescope or the body of the instrument. As with the lenses, the quality of the bubble levels dictate the quality, accuracy, and price of the instrument.

Bubble levels are curved glass vials that contain alcohol or other liquids that are not susceptible to freezing. The bubble inside the vial naturally moves to the highest point under the curved glass. The level is constructed so that the telescope tube's horizontal center line is parallel with the ground at the same time the bubble in the vial is at the highest point in the vial.

There are two basic types of bubble levels: the curved tube level and the circular or bull's eye level. See Figure 2-3. The accuracy and sensitivity of bubble levels depends on the quality of their materials and the degree of curvature of the glass immediately over the bubble.

For example, specifications describe the sensitivity of the curved tube bubble level in terms of the degrees of arc per division. Example: 2 mm per 10 feet of arc.

The divisions or marks on the level tubes on most surveying instruments are 2 mm or .07874 inches apart. Since the tube is in the shape of an arc, lines drawn from adjacent division marks toward the center of the circle would be separated by a small angle. See Figure 2-4. This angle is usually about 10 minutes of arc for most construction levels, but may be as little as 10 seconds of arc for high-precision engineer's levels. The smaller the degrees of arc per division, the greater the radius of the bubble tube arc, and the greater the sensitivity of the bubble tube. Figure 2-4 also shows the means of calculating

(a)

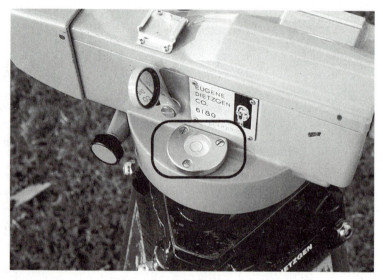

(b)

Figure 2-3.   (a) Tube level; (b) Circular level.

The bubble vial is curved. The divisions are .07874" apart. The segment represents 10' of arc. What is the radius of the circle represented by this segment?

1. Find the circumference of the circle.

$$10' = .16667°$$

$$\frac{.16667°}{360°} = \frac{.07874"}{c}$$

$$.16667c = 28.3464$$

$$c = \frac{28.3464}{.16667}$$

$$c = 170.075"$$

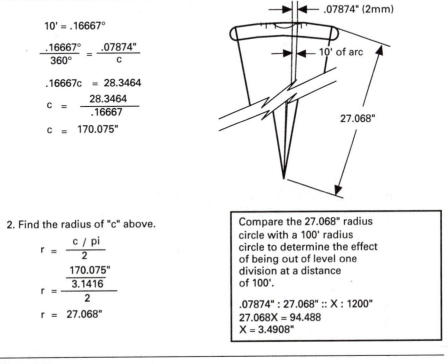

.07874" (2mm)

10' of arc

27.068"

2. Find the radius of "c" above.

$$r = \frac{c \,/\, pi}{2}$$

$$r = \frac{\frac{170.075"}{3.1416}}{2}$$

$$r = 27.068"$$

Compare the 27.068" radius circle with a 100' radius circle to determine the effect of being out of level one division at a distance of 100'.

.07874" : 27.068" :: X : 1200"
27.068X = 94.488
X = 3.4908"

Figure 2-4.

the relative radius and deviation per division per 100 feet when the degrees of arc per division specification is known.

With careful use and practice, accuracy to within one-fourth of a bubble division can easily be obtained with a well-adjusted instrument. This would mean that accuracy to within .0872 inch in 100 feet can reasonably be expected using an instrument equipped with 1 minute of arc per division bubble levels.

Both dumpy and wye levels are capable of being rotated 360 degrees in the horizontal plane. Some have the capability of measuring horizontal angles, but only to about 15 minutes of arc.

Dumpy and wye levels may be purchased or rented from local contractors' or surveyors' supply sources. Occasionally, used levels may be found in second-hand stores or at private sales. Caution should always be exercised in purchasing any used surveying equipment. It may have hidden defects which are not readily apparent to the inexperienced instrument user.

Automatic Levels

Automatic levels have been on the market for several years. These instruments have built-in compensators which cause the level to return to a level setting after being knocked off level by as much as a half bubble on some models. This is particularly helpful in the case of windy conditions and where heavy equipment may cause vibrations which affect the level setting. The term "automatic" may be somewhat misleading as these levels must be manually brought to a level setting before the automatic features take over.

Laser Levels

Laser levels are low-level laser beam transmitters that emit a visible (heliumneon) or an invisible (infrared) beam. The laser beam produced by these construction lasers is concentrated to a diameter of about ⅛ inch at a distance of 200 feet from the transmitter. Figure 2-5 shows a laser level and its detector.

Some laser transmitters are designed to emit their beam only in a straight line. Other transmitters emit a rotating beam much like a lighthouse light beam. Of course, the laser beam rotates much faster and is much more concentrated than a lighthouse beam. Some specially designed laser transmitters can emit both straight-line and rotating beams.

The transmitters for construction lasers are usually equipped for both battery and AC current operation. Most have an automatic self-leveling feature that assures a constant level condition even if the unattended transmitter settles or is otherwise knocked slightly out of level. However, if a major jolt is sustained, the laser will shut itself off until it is re-leveled and reset by the operator.

Many accessory items are designed to allow the transmitter to be mounted in various convenient locations around the job site. However, a standard surveyor's tripod is the most frequently used mounting device for construction lasers. In some cases, laser transmitters are clamped to beams or columns some distance above the floor and other high-traffic areas. Some lasers are designed to be mounted in horizontal pipes with the beam focused on the far end of the pipe. Most of these pipe-laying lasers have the capability of having the degree or percent of grade dialed into them. In this case, the beam will be emitted on the precise angle required for laying the pipe.

Figure 2-5. Laser level, with detector on the left and transmitter on the right. (Photo courtesy of Spectra-Physics.)

The device which intercepts the beam is called a detector. When using a visible-beam laser under normal lighting conditions found in buildings under construction, a small metal or plastic card with a special coating may be used to detect the presence and precise location of the beam. Electronic detectors are used to intercept both visible and invisible beams. These detectors emit both an audible and a visible signal when the beam is intercepted.

There are several advantages to using laser levels in specific situations. The primary advantage is its one-person operating capability. Once the transmitter is set up and activated, there is no need for someone to stay at the transmitter directing the leveling operation and taking readings, as is the case with the conventional telescope level. One field engineer can set up the transmitter, then take a detector to various areas on the job site and intercept the beam.

The field engineer can then compare the elevation of the beam to the required elevation at a specific location on the job site. Simple mathematics (covered later in this chapter) will tell the field engineer whether the elevation needs to be adjusted. As long as the beam is unbroken by obstructions or is not diffused by dusty conditions, the field engineer can rely on it for readings at distances of 300 feet or more from the transmitter.

A secondary advantage to using a laser level is that it may be used by more than one craft at a time. For example, if the height of the beam in relation to the job-site benchmark is known, both plumbers working below floor level and ceiling installers working near the roof can use the same beam to determine the correct elevation for installing both pipes and suspended ceiling grids.

Some types of laser levels are used in utility construction and high-rise construction. These subjects are covered in Chapters 8 and 11, respectively. Additionally, laser detectors are often mounted on construction equipment as an aid to the operator in controlling the depth and slope of ditches and other excavations. In some cases, the detectors are connected to the control systems of heavy earth-moving and paving equipment. These interconnected systems control the equipment's cutting blades and material-spreading surfaces more precisely than they can be controlled by a human operator.

Most laser level detectors can easily be fastened to a standard level rod or they can be fastened to a piece of lumber of similar size. A common practice is to rip an 8-foot 2 × 4 and mount the detector on one of the pieces.

Hand Levels

A hand level is a short (4 or 5 inches long) inexpensive hand-held telescope with a bubble level built into it. The bubble level is visible when sighting through the telescope. If the bubble is centered, the crosshairs of the scope will fall on some object that is at approximately the same elevation as the eye of the field engineer. Figure 2-6(a) shows a person sighting through a hand level.

The hand level is useful when deciding on where to set up a tripod when running a leveling operation. It is often difficult to determine with only the unaided eye just how much a parcel of land is sloping. Nearby structures and excavations may create optical illusions that are particularly bothersome to inexperienced field engineers and to those who have visual acuity problems. Using a hand level to sight on a benchmark, the field engineer can quickly tell if the height of his or her eye is above or below the benchmark. This can be done much faster than setting up a tripod, installing the instrument on the tripod, and leveling the instrument, only to find that, for example, the line of

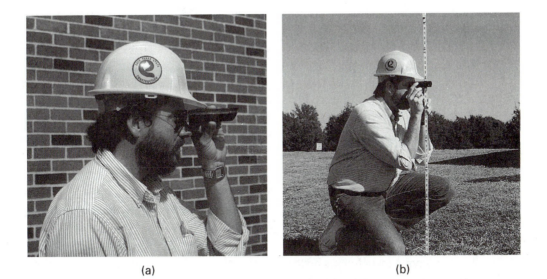

(a)   (b)

Figure 2-6.

sight hits the ground a foot below and in front of the benchmark. In this case, the instrument must be moved uphill and the process repeated until the line of sight is just above the benchmark.

Hand levels may be used in conjunction with a folding rule or a level rod to take quick rough differential elevation readings. Figure 2-6(b) shows how a folding rule and a hand level may be used in this manner.

The hand level is moved up and down the rule until the crosshairs rest on a point of known elevation and the bubble indicates that the tube is level. The reading on the rule directly under the hand level is subtracted from the point of known elevation. The result is the elevation of the ground at the point where the end of the rule is resting. While not extremely accurate, this one-person activity will provide some indications of differential elevations. These indications are sufficient for initial dirt-work operations.

Advanced Technology in Levels

New models of levels are regularly being introduced onto the market. The latest innovation in levels is the use of bar codes as well as numbers on the level rod. The level emits a beam that reads the bar code on the rod. The rod reading is then displayed on a digital readout window on the instrument.

Other Levels

Engineers' levels, transits, theodolites, and electronic instruments may be used for leveling if a dumpy or wye level is not available or if more exacting specifications require a higher degree of accuracy. The same procedures used in setting up a dumpy or wye level are used to set up an engineer's level. The procedures for setting up transits, theodolites, and electronic instruments are slightly different and are covered in subsequent chapters.

Selecting the Tripod

The primary consideration in selecting a tripod is to choose one that is compatible with the instrument selected. The thread design of the tripod-instrument connection and the general structure and stability of the tripod are primary considerations.

Surveying instruments have mounting threads in their bases. These threads must match the threads on the tripod fitting. Thread design is specified by the diameter of the item threaded and by the number of threads per inch or "TPI." There are several thread designs in use today on instruments and tripods. The three most commonly found ones are $1\frac{3}{8}$" × 13 TPI, $\frac{5}{8}$" × 11 TPI, and $3\frac{1}{2}$" × 8 TPI. Figure 2-7 shows tripods with these three types of fittings.

Figure 2-7.   Instrument tripods showing different thread types.

It is the responsibility of the field engineer to ensure that the tripod and instrument threads are compatible and that they are undamaged. Prior to taking a tripod and instrument to the field, any protective covers should be removed and the threads inspected carefully for defects. The instrument and the tripod should be trial mated before leaving the storage area.

The secondary consideration is to determine whether the instrument to be used has an optical plummet and whether it is to be used. An optical plummet is a series of lenses that allow the field engineer to observe the ground immediately under the instrument by looking through an eyepiece on the body of the instrument. This is done for the purpose of positioning the instrument directly over a point without using a plumb bob. If the instrument has an optical plummet and if it is to be used, the tripod center fitting must be open to allow sighting through the plummet. The optical plummet is seldom used in leveling operations. It is covered in detail in Chapter 3.

The type of instrument which is to be used and the terrain must also be considered in tripod selection. Heavier, more sophisticated, and more delicate instruments require heavy tripods to provide support and prevent movement of the instrument. Heavily constructed tripods are more stable, but they are much more cumbersome than lighter-weight tripods. Available features such as adjustable-length legs and level tubes built into the instrument mounting base must be considered as well.

Many times, both metal and wooden tripods are available. Most field engineers will choose wooden tripods with adjustable-length legs. Wooden tripods are less susceptible to dimensional changes caused by temperature variations than are steel or aluminum tripods. Wooden tripods are always chosen if any magnetic-compass work is to be done with the instrument. Of course, the adjustable legs give the field engineer more flexibility in setting up the tripod in less-than-ideal locations.

Selecting the Level Rod

The design and construction of level rods runs from the crude stick with notches carved on it to expensive telescoping rods with vernier-scale targets. Carpenters and other crafts often improvise and build level rods using old tapes or whatever else is at hand. Figure 2-8 shows a "homemade" rod. While the "homemade" rod may suffice in some cases, the professional field engineer needs a more reliable and accurate level rod with which to perform precision leveling operations.

Every manufacturer of surveying equipment offers rods of many different

Figure 2-8. "Homemade" level rod.

designs, each for a specific use. Rods may be purchased with scales marked in feet and inches, feet and tenths of feet, or meters and millimeters. Figure 2-9 shows some of the various markings available. Some rods have both standard and metric scales.

Most rods are numbered from the bottom up, but rods numbered from the top down can also be ordered. Rods may be made of wood or fiberglass, and

Level Rod Faces

Rod faces pictured are approximately one-half (57%) actual size.

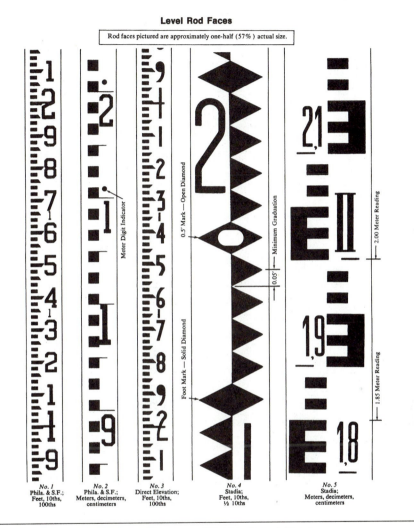

No. 1	No. 2	No. 3	No. 4	No. 5
Phila. & S.F.; Feet, 10ths, 100ths	Phila. & S.F.; Meters, decimeters, centimeters	Direct Elevation; Feet, 10ths, 100ths	Stadia; Feet, 10ths, ½ 10ths	Stadia; Meters, decimeters, centimeters

Figure 2-9.   Level rod designs. (Courtesy of Sokkia Corporation.)

can be purchased in several different quality grades depending on the accuracy of the scales installed on the rods.

Some rods with distinctive designs have been given geographic names, perhaps indicating the area of their origin. For example, the "Chicago" rod is a 12-foot-long three-piece telescoping rod. The "Florida" rod is 10 feet long and is graduated with red and white stripes. "San Francisco" rods are three-piece rods similar to the Chicago rods but are available in several lengths.

The most frequently used rod is the "Philadelphia" rod, or—as it is sometimes called—the "Philly" rod. The Philadelphia rod is a two-piece telescoping rod with scales on both the front and back. It is graduated in feet and tenths of feet. Most are 7 feet long and can be telescoped to 13 feet in length. The Philadelphia rod can be equipped with a target and vernier scale and can be read to an accuracy of one-thousandth of a foot.

2-4   Reading Level Rods and Targets

Reading the Philadelphia rod, with a scale graduated in tenths of feet, is quite different from reading a feet-and-inches scale. It takes some practice, but it must be mastered because much layout work requires the use of elevations expressed in feet and tenths. It is much easier to add and subtract decimals than fractions. Also, converting to inches and back to tenths with each instrument shot increases the chances of making a serious mistake during the leveling procedure.

When preparing for a leveling operation, the decision must be made as to who will read the rod—the instrument operator or the rod holder. This depends on (1) the distance from the instrument to the rod, (2) the power of the instrument telescope, (3) the level of accuracy required, and, to some degree, (4) the personal preference of the crew members.

If the distance over which the rod is to be read is relatively short and the instrument telescope is powerful enough, the instrument operator can easily read the rod directly to within a hundredth of a foot. Longer shots or shots requiring accuracy to thousandths of a foot require the use of a target which can be moved up and down on the rod. Rod targets are equipped with a vernier scale that allows the rod to be read to within one thousandth of a foot. The target is moved up and down the rod until the instrument operator indicates that it is centered in the crosshairs. The rod holder then reads the rod and the target vernier to obtain the elevation.

Figure 2-10 shows a portion of a Philadelphia rod. Note that the black numbers are tenths of a foot and the black lines are hundredths of a foot. The red number is an even foot.

Figure 2-11 shows a target installed on a Philadelphia rod. Note the clamp screw and the vernier scale. Note also that the target is divided into quadrants. These quadrants are usually painted red or orange alternating with white for visibility. The horizontal line between the quadrants is the "0" point which is brought into line with the crosshairs of the instrument.

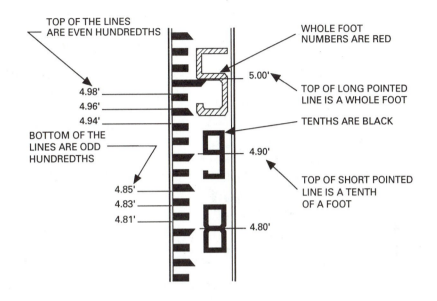

TOP OF THE LINES ARE EVEN HUNDREDTHS

WHOLE FOOT NUMBERS ARE RED

4.98'

4.96'

4.94'

5.00'

TOP OF LONG POINTED LINE IS A WHOLE FOOT

TENTHS ARE BLACK

BOTTOM OF THE LINES ARE ODD HUNDREDTHS

4.90'

TOP OF SHORT POINTED LINE IS A TENTH OF A FOOT

4.85'

4.83'

4.81'

4.80'

PHILADELPHIA ROD MARKINGS

Figure 2-10.   Level rod face. (Illustration by Dale Barnes.)

Figure 2-11.

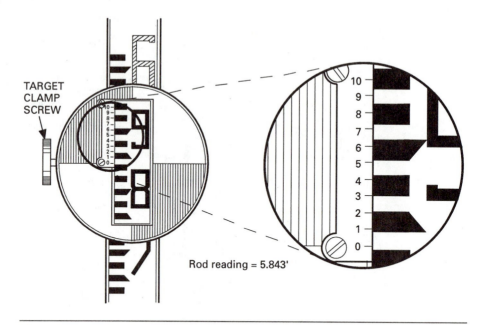

Rod reading = 5.843'

Figure 2-12.   (Illustration by Dale Barnes.)

Figure 2-12 is a line drawing of a rod and target. With the target in its present position, this rod and target reading is 5.843 ft.

The following procedure explains how the rod and target vernier are read and how the 5.843 ft reading was obtained.

STEP BY STEP APPROACH

Step 1. The rod holder should loosen the target clamp screw enough so that the target will slide easily on the rod, but will not fall when released. At the direction of the instrument operator, the rod holder moves the target up or down on the rod until the instrument operator indicates that the target is centered in the crosshairs. The rod holder should then tighten the target clamp screw snugly to lock the target in place on the rod.

Step 2. The rod holder should then observe the first red number which appears on the rod below the target. This is the even-foot indicator. In the case of the example in Figure 2-12, there is no red number shown below the target, but the red number 6 appears just above the target. Therefore, the first red number below the target is a 5. 5 ft would be recorded as the even foot rod reading.

Step 3. The rod holder then observes the rod to determine the first black number which appears on the rod below the target center. In Figure 2-12, this figure is an 8. This is the tenth-of-a-foot value. 8 is recorded as .8 and is added to the 5 ft from step 2 above. The rod reading to this point is 5.8 ft.

Step 4. The black lines are counted next. Beginning at the line directly opposite the black 8 on the rod and counting upward, two black lines are counted before the 0 line on the vernier scale is reached. These two black lines represent 4 hundredths of a foot. The .04 is added to the 5.8 ft that was derived above. The rod reading to this point is 5.84 ft.

Step 5. Once the 0 line on the vernier is reached, the final number is taken from the vernier. This number represents a thousandth of a foot. To read the vernier, the vernier scale is followed upward until one of the lines on the vernier scale lines up perfectly with the edge of one of the black lines. In Figure 2-12, this occurs at the #3 vernier line. The 3 is then recorded as .003 and added to the 5.84 ft from step 4 above. The final rod reading is 5.843 ft.

Step 6. Review the reading. Check that the proper red number and the proper first black number were recorded. Re-count the black. Re-check the line on the vernier which appears to be lined up with the edge of a black line. Check the addition and decimal-point placement. Record the reading in the field book in the proper location.

High Rod

A Philadelphia rod can be extended or telescoped for additional height as needed. A 7-foot rod can be extended to a height of 13 feet. The practice of using an extended rod is called "high rod."

There are two ways to read high rod. If the rod is to be read directly by the instrument operator, the rod must be extended to its full height and locked in the fully extended position. The side of the rod with the scale that has the numbers reading from bottom to top is oriented toward the instrument and read in the conventional manner. The degree of accuracy to which the rod can be read depends on how close to the rod the instrument is located and how powerful the telescope on the instrument is. No vernier scale is used, so the maximum degree of accuracy possible (without estimating) is to the hundredth of a foot.

If greater accuracy is required or if the rod is too far away for the instru-

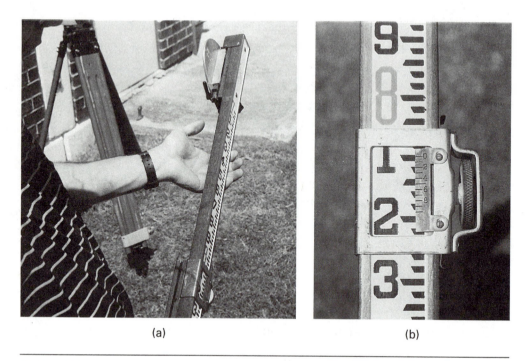

(a)   (b)

Figure 2-13.

ment operator to read, a slightly different technique is used. In this procedure, the target is slid to the very top of the rod and the 0 line on the vernier scale is aligned with the 7-foot mark on the rod. The target is clamped in this position. Figure 2-13(a) shows a target clamped in the high rod position and the rod partially extended. Next, the rod is extended upward until the target is centered in the crosshairs. The rod extension is clamped and the rod holder reads the rod clamp vernier on the back of the rod in the same manner as the target vernier is read. Figure 2-13(b) shows the rod clamp vernier scale.

2-5   Transporting and Moving Surveying Equipment

Surveying equipment can easily be damaged during transportation to and from the job site or while being moved from point to point on the job site. The following is a list of general rules relating to transporting and moving surveying equipment.

1. Keep the instrument in its carrying case when not mounted on a tripod. Keep all protective covers on lenses, threads, or other parts so equipped. Do not allow instruments to be transported unsecured in the cargo area of trucks or in the trunks of automobiles. Protect rods and tripods from shifting cargo.

2. Take care when moving instruments from one environment to another. Moving instruments too quickly between warm offices and cold job sites may cause condensation to form on parts of the instrument and may cause fogging or condensation on the telescope lens. Also, some parts of the instrument may bind due to the different expansion ratios of the various materials. The leveling screws on four-screw instruments should be loosened when the instrument is subject to swings in temperature extremes.

3. Instruments should always be kept dry. It is a good practice to keep a piece of plastic (a clean garbage bag works well) and some string or tape in the instrument's carrying case. If the instrument is set up on the job site and must remain temporarily in wet or dusty conditions, the plastic can be tied or taped around the instrument until such time as the instrument can be moved or conditions improve. However, should an instrument get wet, do not store the instrument until it has been completely dried out. A heat lamp may be used to aid in drying the instrument, but do not let the instrument get so hot that it cannot be handled with bare hands.

4. Tripods should be moved or transported with their legs folded tightly together and secured. Most tripods have a belt or strap attached to one leg which is designed to be used to secure all three legs together in a tight bundle. Tripods should be stored in a dry, secure place in an upright position. The legs on extendible-leg tripods should be fully retracted when stored. As with instruments, all thread protectors should remain in place when the tripod is not in use.

5. When moving from one position to another on the job site, the instrument and tripod may be transported as a unit. The legs of the tripod are gathered into a tight bundle and the tripod, with the instrument attached, is carried over the shoulder or under the arm. Care must be taken not to bump the instrument into protruding items such as pipes or beams.

6. Level rods should be transported and stored in the flat position to prevent warpage. Temporary storage in the vertical position is acceptable, but should be kept to a minimum. Telescoping or extendible rods

should be fully retracted for storage or movement. A fully retracted rod may be over 7 feet long, so care must be taken not to bump into things while moving the rod from one position to another. **Never** allow the rod (even a wooden one) to come into contact with overhead power lines.

 NOTE

Convenient carrying and storage cases for rods can be made of plastic (PVC) pipe sections. Carrying handles can be attached with hose clamps or duct tape. Remove the cap from the case if the rod needs to dry out during storage.

2-6   Setting up the Equipment

Setting up the Tripod

Choosing the location for setting up the tripod for leveling operations is dependent upon job requirements, benchmark locations, and the visibility of the various points where sightings must be taken. Although each job is somewhat different, several common rules that should be observed no matter where the tripod is set up are listed below.

1. Set up the tripod in a location where there is a minimum of foot and equipment traffic. Try to locate a point approximately halfway between the benchmark and the second rod setup point.
2. Set the tripod at an elevation where the instrument telescope will be **above** the benchmark, but not above a level rod set on the benchmark. Use a hand level to aid in selecting such an elevation.
3. Set the tripod legs about 3 feet apart. Adjust the length and location of the tripod legs until the instrument mounting base appears level and is at a convenient working height. Be sure to allow for the height of the instrument to be used. Move around the tripod and sight across the instrument mounting base. Sight on the horizon or on adjacent structures that have level areas exposed. Re-adjust the tripod legs to get the instrument mounting base as level as possible. **Note:** A short level may be used here. Also, some tripods have levels built into the bases.

    After leveling, push the lower points of the tripod legs slightly but firmly into the ground. If the legs are pushed in a uniform distance, the instrument base will remain close to level. The height above ground should be convenient for sighting through the instrument telescope.

4. If wind or other disturbances are a problem, sandbags, cinder blocks, or other weights may be necessary to keep the tripod from being over-turned. But, do not hang weights from the tripod base or legs! Run a cord loosely from the tripod leg to the weight. The weight is just a safety measure. A weight hung from the leg of the tripod may tend to pull the leg downward into the ground, bend the leg, or otherwise disturb the level position of the instrument.

NOTE

If the tripod must be set up on concrete or some other hard, slick sur-face, the legs of the tripod may slip, causing the instrument to fall. Some tripods have short pieces of chain attached to the legs to hold them in position. If necessary, a triangle approximately 3 feet on a side may be made of scrap wood and the tripod legs set inside the triangle. Another alternative is to use a piece of scrap carpet to set the tripod on. If using a transit or a theodolite, a circle may be cut from the center of the carpet if the instrument needs to be centered over a point on the floor or on a line.

Setting up the Leveling Instrument

Once the tripod is properly set up, remove the instrument from its case. Re-move any thread protectors and prepare the instrument for mounting on the tripod. The instrument should be handled by the frame or grasped at the center of the telescope. Place the instrument on the top of the tripod and engage the screw threads. *Do not release your grip on the instrument until you are posi-tive that the tripod and instrument threads are fully engaged and that the instrument is secured on the tripod.*

The screw fittings should just be snug and should not be over-tightened. Once the instrument is secured to the tripod, the instrument is ready to be leveled.

NOTE

Most instruments have a sunshade provided for the object lens. On some instruments, it is permanently attached and must be pulled forward into position. On older instruments, the shade is kept in the instrument case and must be installed each time the instrument is set up. ***Always use the shade.*** It protects the object lens and may prevent severe damage to the instrument if it is accidentaily knocked over.

There are two types of instruments currently in general use. One has three leveling screws arranged in a triangle, and the other has four leveling screws arranged in a square. The four-screw instruments have been in use longer than three-screw instruments, but three-screw instruments are gaining in popularity. Most laser levels are of the three-screw variety.

Leveling a Four-Screw Instrument

The following is the procedure for leveling four-screw instruments.

1. Release any rotation locks on the instrument and check for free rotation in the horizontal plane.
2. Rotate the telescope until it rests directly over two leveling screws. See Figure 2-14(a).

Leveling a four-screw instrument

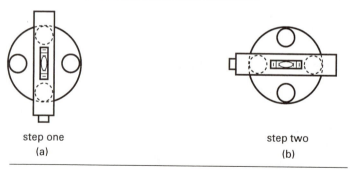

step one
(a)

step two
(b)

Leveling a three-screw instrument

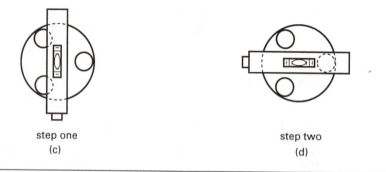

step one
(c)

step two
(d)

Figure 2-14.

3. Grasp the leveling screw knurled wheels that are directly under the telescope with the thumbs and forefingers. See Figure 2-15(a).

4. Look at the bubble level. The bubble in the vial will follow the left thumb. If the bubble needs to be moved to the left to be centered, the left thumb would "pull" the leveling screw wheel to the left. (Clockwise rotation, as viewed from the top.) At the same time, the right thumb would "pull" the right wheel to the right. (Counterclockwise.)

5. If the bubble needs to be moved to the right to be centered, the left thumb would "push" the left leveling wheel to the right. The right thumb would "push" the right leveling wheel to the left. See Figure 2-15(c).

6. When the bubble is leveled across the first set of screws and the screws are both snug (not tight), rotate the telescope 90 degrees and center it over the other set of leveling screws. See Figure 2-14(b).

7. Repeat steps 3 through 5 until the bubble is centered in the vial.

NOTE

The rule of "thumbs" in leveling dictates that (1) The bubble follows the left thumb, and (2) both thumbs move inward together or both thumbs move outward together when grasping the leveling wheels. This means that the leveling screws are being turned in opposite directions. The screws should both remain snug during the operation. It takes some practice to learn to turn each screw just enough to keep it snug while the other is turned in the opposite direction.

---

8. Rotate the telescope slowly through 360 degrees of rotation, stopping occasionally to observe the position of the bubble in the level vial. If the bubble does not remain centered, repeat steps 3 through 7. If the bubble still does not remain centered when the telescope is rotated 180 degrees after being leveled repeatedly, the level vials are out of adjustment. Consult the instrument manufacturer's owner's manual for instructions on how to adjust the level vials, or take the instrument to an authorized repair center for adjustment.

Instruments can be used when the bubble is not centered in the vial. If the bubble is slightly off center and stays in the same position throughout 360 degrees of rotation of the instrument, the instrument is level.

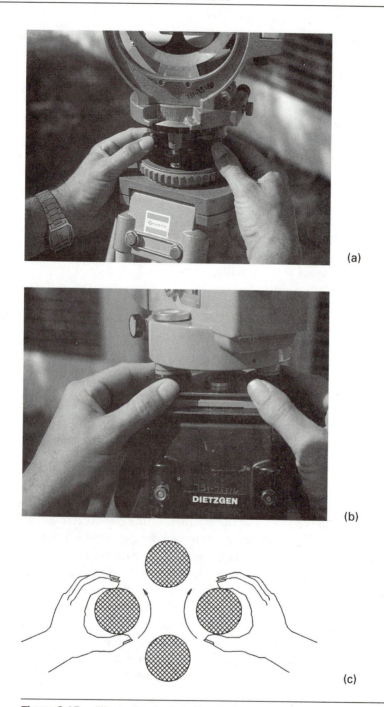

(a)

(b)

(c)

Figure 2-15.   (Illustration by Dale Barnes.)

CAUTION

If a four-screw instrument is supported by a center bearing, adjustment of the leveling screws during a leveling operation will not change the height of the instrument itself. If the instrument is supported only by the leveling screws and does not have a center support, any adjustment of the leveling screws during the leveling operation will change the instrument height and will adversely affect the outcome of the leveling operation.

Leveling a Three-Screw Instrument

The following is the procedure for leveling three-screw instruments.

1. Release any rotation locks on the instrument and check for free rotation in the horizontal plane.
2. Rotate the telescope until it is as nearly over two of the leveling screws as possible. See Figure 2-14(c). Observe the left thumb rule from step 4 above. Continue to rotate the leveling screw wheels until the bubble is centered in the level vial.
3. Rotate the telescope 90 degrees until it is centered over one leveling screw and between the other two. See Figure 2-14(d). Rotate the screw which is directly under the telescope (observing the left thumb rule) until the bubble is centered in the level vial. See Figure 2-15(b).
4. Rotate the telescope slowly through 360 degrees of rotation, stopping occasionally to observe the position of the bubble in the level vial. If the bubble does not remain centered, repeat steps 2 and 3. If the bubble still does not remain centered when the telescope is rotated 180 degrees after being leveled thoroughly, the level vials are out of adjustment. Consult the instrument manufacturer's owner's manual for instructions on how to adjust the level vials, or take the instrument to an authorized repair center for adjustment.

NOTE

Some instruments have two level vials mounted on the body of the instrument. One vial will be mounted at 90 degrees to the other. To level, center one vial over opposing leveling screws, just as with a single-vial instrument. The bubble in the first vial is centered, then the other set of leveling screws (or single screw on three-screw instruments) is used to center the bubble in the second vial. Subsequent procedures are identical to those used with other instruments.

Instruments with bull's eye levels are leveled in much the same manner as those equipped with tube levels. The bubble is "pushed" toward the center of the vial with the left thumb.

CAUTION

Three-screw instruments do not have a center support, therefore any adjustment of the leveling screws during a leveling operation will change the height of the instrument. This will adversely affect the outcome of the leveling operation.

NOTE

Once the instrument has been mounted, centered, and leveled, the tripod should not be touched! Inexperienced field engineers have the tendency to grasp the tripod to steady themselves while looking through the telescope. This practice will often disturb the legs of the tripod enough to throw the instrument out of level. See Figure 2-16.

Setting up the Level Rod

The proper setup of the level rod is as important as the proper setup of the leveling instrument. In some cases, it takes more skill to properly set and read the level rod than it takes to operate the leveling instrument. Since the Philadelphia rod is the most commonly used level rod, it will be used as an example in this section. All other rods are similar and some are easier to use, though not as versatile as the Philadelphia rod.

Selecting the proper position on the job site to set up the level rod is dictated by what needs to be accomplished in a given situation. If a simple one-shot transfer of an elevation from a benchmark to another point on the job site is all that is required, the rod may be set up in any convenient location near the point where the new elevation must be established. There is no need to worry about the setup point being eradicated by future construction activity. If necessary, the elevation could be re-established quite easily. If, however, the level rod setup is a part of a more extensive leveling operation, one in which permanent or semipermanent benchmarks must be established, more care must be taken in the general setup procedure.

(a)

(b)

(c)

Figure 2-16.   (a) and (b) Incorrect ways and (c) correct way of working with an instrument mounted on a tripod.

The following is a list of general rules for level rod setup. Of course, there are often circumstances in which all these rules cannot be followed. Special arrangements must often be made in such cases.

1. The rod holder must confer with the instrument operator about exactly what is to be done and on which communication systems and signals are to be used. For instance, it must be decided whether high rod or target vernier is to be used.

2. Select a position out of the way of vehicle and foot traffic if possible. If it is necessary to set up on or near roadways, proper safety precautions must be observed. Check for overhead obstructions. Do not set up where the rod will come close to overhead wires.

3. Select a position where the rod can be set on firm ground and where the person holding the rod can maintain firm footing.

4. Select a position which is no farther from the instrument than the instrument is from the existing benchmark.

5. If the rod position is to become a reference point or benchmark for future work, set a hub stake and guard stakes with the proper notations and flags on them. (See Section 2-8.)

6. Determine that the instrument operator can see the rod. Does the rod need to be extended?

7. Take a position behind the rod with the face of the rod oriented toward the instrument. Establish a firm but comfortable stance, holding the rod in a vertical position by checking the rod level if one is present. Use range poles or other braces in windy conditions. See Figure 2-17.

8. Observe the instrument operator for signals or listen for verbal instructions. Do not move the rod until instructed to do so by the instrument operator. (This is particularly important if the instrument is to be moved and the rod is to be sighted from the second instrument position.)

9. Record rod readings as required.

An experienced rod holder can anticipate the directions of the instrument operator and can contribute immeasurably to the efficiency and accuracy of the leveling operation. The position of rod holder on a field engineering crew is usually the training position for more responsible positions, such as instrument operator and field party chief. Most chief field engineers have served some time as rod holders, thereby gaining experience which allowed them to advance in their profession.

Figure 2-17. Method of using range poles to steady a level rod.

2-7 Differential Leveling

The process of transferring elevations and establishing new benchmarks is called differential leveling. This simply means measuring the "difference" in elevations between two or more points. These points may be quite close together or may be miles apart. Even though two benchmarks may be close together, there may not be a clear line of sight between them, and there may or may not be a third point where both are visible. It is the responsibility of the field engineer to measure the differences accurately between benchmarks and to establish new benchmarks as required.

To measure the difference in elevations between benchmarks, the field engineer must go through several precise steps. Each step must be carefully documented. Of course, the more steps required to reach the objective, the more chance for error.

The following is a step-by-step procedure for establishing or checking the elevation for the top of a concrete grade beam prior to placing the concrete in the forms. This example uses a previously established job-site master benchmark and one intermediate benchmark (called a "turning point" or "TP") established by the field engineer. This example is typical of the work performed by field engineers on a daily basis. Refer to Figure 2-18 while studying the step-by-step procedure below.

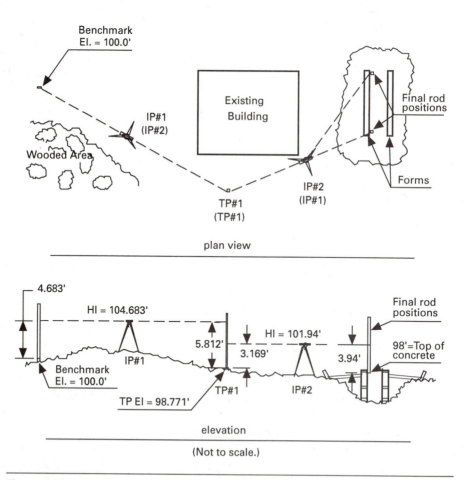

Figure 2-18.   (Not to scale.)

STEP BY STEP APPROACH

Personnel required: Instrument operator, rod holder, and note taker.

Tools required: Proper leveling instrument for this job, level rod, field notebook, safety gear as required, field engineer's tool box, hub and guard stakes.

NOTE

If a laser level is to be used, one field engineer may be able to do this job alone. The procedure would be only slightly different.

Step 1. Consult the blueprints to determine the location and elevation of the job-site master benchmark and the elevation of the top of the grade beam. Note the difference. Is the beam to be above or below the benchmark? Check the prints for the design of the overall structure. Plan a preliminary route (traverse) for the leveling operation. Try to pick tripod setup points that will be halfway between benchmarks and TPs. Check with the project manager as to the progress of the project as it relates to your proposed route and to see if any hazardous activities are planned which will affect the field engineer's work.

Step 2. Check the equipment to be used. Is it complete, compatible, and appropriate for this job?

Step 3. Proceed to the job-site master benchmark. Observe the benchmark and the surrounding area closely. Does the benchmark appear to be undisturbed? Is there any indication that it may have been run over by heavy equipment or has heavy equipment passed closely enough to displace the benchmark? Are the guard stakes undamaged and are the notations on them clear? If the master benchmark appears to have been disturbed, notify the project manager. It may be necessary to re-establish the benchmark or to check it for accuracy. This may be done by checking the elevation of several completed parts of the structure against the original elevation of the benchmark. There may be a second benchmark nearby that can be used as a check as well.

Step 4. Visit the proposed locations of the intermediate benchmark (TP), the tripod setup points, and the final elevation location on the concrete forms. Does each point have the necessary visibility and elevation requirements? Using a hand level, determine that the instrument setup points are at an elevation where the benchmark and the TP can be seen.

NOTE

It is highly recommended that the inexperienced field engineer sketch out the field problem, noting the TP points, the instrument setup points, and the various elevations. Both profile and plan views of the field problem should be drawn. This aids tremendously in visualizing the field problem.

Step 5. Have the rod holder set the rod on top of the job-site benchmark and hold it plumb. The field engineer should proceed to the first tripod setup point. The tripod and level are assembled, leveled, and sighted on the rod. This is called the "backsight" because the sight direction is "back" toward the level rod from an advanced tripod position.

Step 6. Plumb the rod and take the reading. The rod holder should hold the rod plumb using a rod level. If no rod level is available, the instrument operator should determine that the rod is parallel with the vertical crosshair of the instrument and give the proper signal to plumb the rod if necessary. Signals are discussed in Section 2-10. When the rod is plumb with the vertical crosshair, the signal should be given to the rod holder to gently rock the rod a few inches toward the instrument and back away from the instrument by a like amount. The top of a 7-ft rod should move about a foot and a half during this operation. This practice tends to negate any out-of-plumb condition in the rod/instrument axis. The lowest elevation observed during this toward-and-away-from rocking of the rod should be recorded.

NOTE

When extreme accuracy is required, the level is rotated, re-leveled, and another series of readings is taken. The average of several sightings is recorded as the backsight elevation. Also, the leveling route or "traverse" may be run backwards to check accuracy.

If the rod holder is to take the reading, the instrument operator should have the rod holder adjust the target on the rod to the lowest point observed during the rocking motion. Of course, this rocking procedure can be omitted if the rod is plumbed with a rod level. Record the observed readings in the field book. The benchmark elevation plus the rod reading at the benchmark is called the "height of the instrument" or "HI."

Step 7. Establish the secondary benchmark. The rod holder moves to the position previously visited and determined to be the point of the secondary benchmark. This point is designated as TP-1. If this point is on concrete, spray paint or chalk is used to mark the exact spot where the rod is to be set. If the point is in soil, a hub stake and guard stakes are put in place. The proper notation is printed on the guard stakes and the proper color of surveyor's ribbon is attached to the guard stakes. Figure 2-19 shows a hub stake with guard stakes and ribbons. The rod is set on the hub stake or the spot marked on the concrete and then the rod reading is taken. The rod reading methods explained in step 6 may be repeated as required.

(a)   (b)

Figure 2-19. Hub and guard stakes.

NOTE

This second reading is called the "foresight" or "frontsight" because it is taken going "forward" from the instrument location. The foresight reading is subtracted from the "HI." This gives the elevation at this turning point (TP-1).

Step 8. Move the instrument to the second instrument point. The rod holder should stay in position at TP-1, preferably keeping the rod in place on the marked spot. The instrument operator should pick up the level and tripod and move to the second instrument point. The instrument and tripod should be set up, leveled, and focused on the level rod. The rod readings are taken as before and recorded.

Step 9. Take the last rod reading. After the rod readings at TP-1 are complete, the rod holder should move to the final objective, in this case, the concrete forms. At this point, some simple mathematics are used to determine the exact foresight elevation needed to have the base of the rod set at a height corresponding to the top of the proposed grade beam. To do this, simply subtract the desired elevation from the HI on the last foresight. The rod target is set at this point on the rod. Next, the rod is lowered into the forms until the target is centered on the horizontal crosshairs of the level. The base of the rod is at the level of the top of the proposed grade beam. A pencil mark drawn on the inside of the forms at the elevation where the base of the rod rests shows the level to which the concrete should rise in the forms during placement.

Procedure Notes

The general practice is to do a rough preliminary leveling operation for the purpose of setting forms and reinforcing steel. Then a second and more accurate leveling operation is done to establish the final level of the concrete. Marks at the base of the level rod placed on the inside of the concrete forms will be connected by popping a chalk line from one mark to another. This chalk line should be popped before form oil is sprayed on the forms. The chalk line will be destroyed by the wet concrete when it starts to flow into the forms, so some other indicator must be installed at the elevation of the chalk line to guide the concrete placement crew. A row of nails, a chamfer strip, or something suspended from the top of the forms will be installed to show the concrete placement crew how much concrete to put in the forms.

Step 10. Complete the job. After the final marks have been placed inside the concrete forms, the best practice would be to "close the traverse." This simply means to continue the leveling process using another route to get back to the original benchmark. This may involve establishing one or two more temporary benchmarks in the same manner in which the first one was established. Usually the same care in setting the hubs and guards is not taken because these hubs will be removed after the traverse is closed.

Once the level traverse is traced back to the original benchmark, the difference between the final elevation and the original elevation should be quite close. Just how close depends on the distances covered, the quality of the instruments used, and the skill of the field engineering crew. Variations of as little as ⅛" to ¼" can routinely be obtained in a closed traverse leveling operation which uses six TP points 500 feet apart.

If the traverse cannot be closed by another route, some field engineers will re-trace the traverse back to the original benchmark. This provides a double check on the original process.

Complete the paperwork by completing the field notes and the sketches. The field notes should contain information on the readings and calculations as well as the names of the crew members, weather conditions, page number of the blueprints used, and references to any unusual conditions encountered. The sketch of the traverse should show the primary benchmark, the route taken, and the location of all tripod and TP points. Some field engineers will have the rod holder check the field notes and initial them before they are filed. Figure 2-20 shows how the field notes might look for this leveling operation.

Next, inventory the equipment and store it in accordance with recommended practices.

After the field notes have been filed and the equipment stored, the field engineer should report to the project manager that the leveling assignment has been completed and that the field engineering crew is ready for another assignment.

2-8   Math Formulas

The mathematics involved in leveling processes usually consist of simple addition and subtraction. The only difficulty comes in keeping track of where one is in the operation and what should be added or subtracted from what. Good field notes are essential to keep mistakes to a minimum.

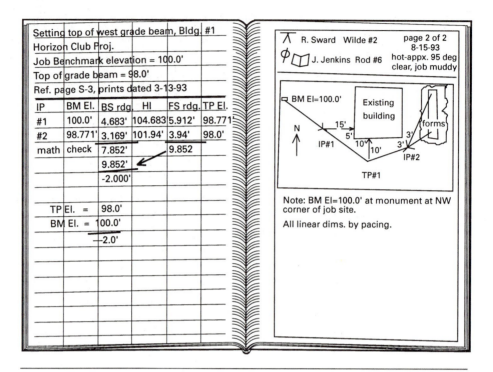

Figure 2-20.

Figure 2-20 shows the field notes from a typical leveling operation. Note that the route of the leveling traverse is sketched on the right page and that the locations of the tripod setup points and benchmarks are clearly indicated.

The benchmark elevation is recorded prior to beginning work. Rod readings are recorded as each step progresses. After the traverse is complete, the backsight and frontsight columns are summed.

The frontsight total is subtracted from the backsight total. This may produce a negative number if the final elevation is lower than the beginning benchmark. This is the case in Figure 2-20. Next, the final elevation is subtracted from the beginning benchmark. This figure and the figure obtained by subtracting the frontsights from the backsights should be the same. If it is not, review the mathematics involved.

The principal thing to remember is to "add the backsight and subtract the foresight." A good way to remember this is to picture the traverse in profile. The elevation route is from the benchmark "up" the rod (adding), through the telescope, and "down" the rod (subtracting) to the next point (TP). This is expressed as

benchmark elevation + backsight – foresight = TP elevation

and also as

benchmark + backsight = HI elevation

It is considered good practice to sit down in a quiet place after the traverse has been completed and review the entries in the field book.

2-9   Locating and Transferring Benchmarks

Benchmarks are considered to be relatively permanent elevation markers. These markers usually consist of brass disks or other durable items set in concrete or stone. The locations for these benchmarks are chosen carefully; they must be accessible, yet protected and stable.

Perhaps the most widely used primary benchmarks are those installed by government agencies. The U.S. Department of the Interior Geological Survey has established many benchmarks throughout the country. Also, many local government agencies have established their own benchmarks for local elevation control. These local control points are sometimes called "datum points" or "datum elevations." Sometimes both U.S. and local government benchmarks are used as reference points in establishing construction project benchmarks.

If a government agency benchmark is to be used as a reference point to establish elevation control for a construction project, it is usually necessary to establish a secondary benchmark on the job site. It is rare that a government benchmark is close enough or placed conveniently enough for use on a particular job site.

Locating and transferring benchmarks is similar to a job-site differential leveling operation, as discussed in Section 2 -7. The only difference is in the level of accuracy required and the method of establishing the job-site control point.

The level of accuracy required for establishing project benchmarks may be as high as 1:50,000 (approximately .1 inch in 500 feet). Obtaining this level of accuracy requires the use of expensive precision instruments with very sensitive level vials. Repeated sightings on invar rods and corrections for light refraction and earth curvature may be required as well. If this is the case, a professional surveyor is usually employed to set the job-site benchmark.

Of course, many projects do not require this level of accuracy, so many construction companies set their own job-site benchmarks using their own field engineering crews. The major differences in procedure for setting a job-site benchmark and in other differential leveling operations is that more care is taken to prevent errors and the mark itself is mounted or established in a more permanent fashion.

The job-site benchmark location is usually established first. The location must be accessible and protected at the same time. Not only must the location be accessible at the beginning of the job, it must be accessible throughout the job. It must not be shielded from view by office trailers or be covered with excavated material or any other obstruction. It must be in a position where it will not be disturbed by landscaping, or by utility, road, or parking lot construction. It should be visible from as nearly one hundred percent of the job-site as possible.

The project manager should be consulted on the placement of the job-site benchmark. He or she is in charge of the method and sequence of construction and will know if certain areas of the site are going to be used for storage, temporary offices, and so forth. It may be necessary to have clearing and demolition done or dirt and debris moved before the benchmark placement can proceed.

The site for the job-site benchmark must be stable as well as accessible. It cannot be in a low, swampy area or in an area where shifting of the soil will be a problem. Care must be taken not to locate the benchmark over an area where tunneling is to be done because surface subsidence may occur. The site must also be protected from vandals.

Once the site is chosen, a common practice is to excavate a hole 12 inches to 18 inches square and about 2 feet deep. A wood form about 4 inch tall is set around the edge of the hole and concrete is poured level with the top of the form. (Some field engineers prefer to dig a round hole with a post-hole digger.) Reinforcing steel or wire is optional, but usually not necessary.

After pouring, a metal marker is imbedded in the wet concrete. Brass disks especially designed for use as benchmarks are available from surveyors' supply stores. These disks have stems on them which reach several inches into the concrete to prevent easy removal. However, many field engineers prefer to simply use a piece of concrete reinforcing steel approximately 12 inches long set vertically in the center of the concrete with the tip sticking up out of the concrete about $\frac{1}{4}$ inch.

When the concrete has cured, the "monument" (as the concrete marker is called) is "protected" by guard stakes or posts, or by a low fence if necessary. Plenty of ribbon and paint is used to warn equipment operators and others away from the area.

After the monument is set, the field engineering crew is ready to begin the transfer work. The most accurate instruments available and rod targets with vernier scales are used. The instruments are positioned as nearly as possible halfway between turning points (TPs).

The benchmark transfer operation proceeds just like any differential leveling operation but with more care. In some cases, repeated readings are taken and the results averaged. In other cases, readings are taken, the telescope is rotated 180°, the reading is taken again, and the results are averaged. In some cases, another field engineering crew is asked to run the same traverse. The notes taken by the two crews are compared and discrepancies worked out. The field engineer quickly learns which of these methods works best with a given instrument in a particular situation.

Once the elevation of the top of the brass plate or rebar in the monument is determined, the figure may be stamped into the brass plate, chiseled into the concrete, or noted on a guard stake. The benchmark elevation is recorded in the field book and is reported to the project manager. It may also be noted on the project prints and be recorded in the "as-built" prints after the job is completed. In many cases, the project manager will relay the information concerning the monument to various subcontractors and to the project architect.

NOTE

Some field engineers prefer to set a longer piece of rebar in the monument, run the differential leveling operation, and then cut the rebar off at an even-foot mark.

2-10   Additional and Advanced Techniques for Leveling Operations

Hand Signals and Radio Communications

Because field measurements and leveling crew members are usually separated by some distance and must work in noisy areas, hand and arm signals are often used to convey instructions between crew members. Figure 2-21 shows the standard signals used by surveyors and field engineers. Although the signals shown in Figure 2-21 are the recommended standard ones, field engineering crews often make up special signals of their own and use them to advantage.

Of course, walkie-talkie type radios are being used more frequently on today's job sites. These lightweight and relatively inexpensive radios have improved communications markedly. This has resulted in a reduction in the number of mistakes being made and in a great deal of time being saved.

If radios are to be used, try to obtain the variety which is voice activated.

INSTRUMENT OPERATOR HAND SIGNALS

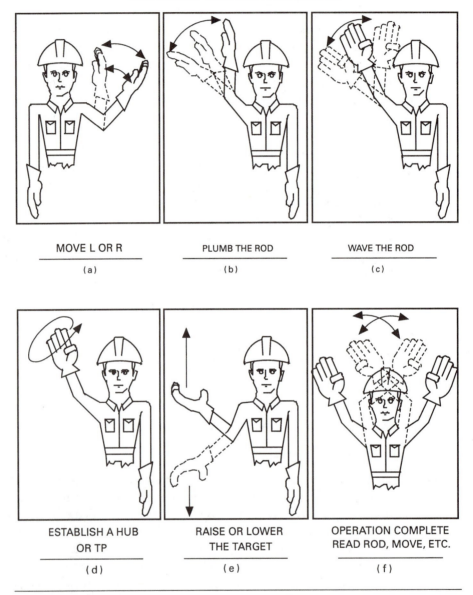

MOVE L OR R	PLUMB THE ROD	WAVE THE ROD
(a)	(b)	(c)

ESTABLISH A HUB OR TP	RAISE OR LOWER THE TARGET	OPERATION COMPLETE READ ROD, MOVE, ETC.
(d)	(e)	(f)

Figure 2-21.   Surveying hand signals. (Illustration by Dale Barnes.)

They do not require that a button be pushed in order to transmit, and they can be carried on a tool belt. Most can be equipped with a headset and a "boom" microphone. These features make possible hands-off operation.

It is also advantageous if the radio has more than one channel available. It is easier to carry on the field engineering operations if one channel is dedicated to this activity. General job-site communications should be on a separate channel.

Remember, it is imperative that the radio battery packs be fully charged prior to beginning work.

Inverted Rod

In some cases, the field engineer may find it advantageous to use the rod in the inverted (upside down) position. This procedure may be used in checking the elevation of overhead forms, such as those for concrete ceilings or elevated concrete beams. This procedure may also be used to check grid elevation for suspended ceilings and elevations of piping or duct work that is ceiling mounted.

Inverted rod is usually used to check installed components and is rarely used to give an elevation from which components are installed. However, the final adjustments of concrete ceiling forms may be done much more easily from below using inverted rod.

The field engineer must use caution when using inverted rod. It must be remembered that the numbers are being read in reverse. It helps to stop and draw a sketch of the problem with dimensions before proceeding.

Trigonometric Leveling

Trigonometric leveling is sometimes used when a backsight is not possible or when only differential elevations are needed which are not referenced to a benchmark. For example, the height of a building may be determined by setting up an instrument some distance away and sighting on the base of the building. The vertical angle is recorded.

Next, the telescope is rotated up and the top of the building is sighted. The vertical angle is again recorded. Once the distance from the instrument is measured, trigonometry can be used to determine the height of the building.

This procedure requires an instrument capable of measuring vertical angles. These instruments are discussed in Chapter 3.

Three-wire Leveling

The technique of three-wire leveling is no more than taking three rod readings at the same time. Readings at both stadia hairs and at the cross hair are observed and recorded. The three readings are averaged and the average is used as the elevation of the backsight or frontsight as the case may be.

This procedure usually produces a fourth digit to the right of the decimal in elevations. This increases accuracy to some degree, but the chances of an error in taking the readings is increased as well. This procedure is seldom used in construction layout work or in setting job-site benchmarks because the level of accuracy is not necessary.

Peg Test (or Collimation Test)

Sometimes the bubble vials on levels and other instruments get out of adjustment. That is, the bubble is not centered in the tube when the line of sight through the telescope is horizontal. The experienced field engineer will periodically check the levels and other instruments to see if this condition exists.

The procedure to check for the out-of-adjustment bubble is called the "peg" test. To perform this test, follow the procedures listed below.

1. In a fairly level and clear area, set two hub stakes 200 ft apart. By measure, set up the suspect level halfway between the two hubs. The instrument doesn't have to be precisely on the line between the hubs, but it must be set up an equal distance from each hub. Have an assistant hold a rod on the hubs and take several frontsights and backsights readings. Average the readings and subtract the averages to determine the precise difference in elevation between the hubs.

2. Move the instrument to a position within a foot of one hub. Set up and level the instrument. Have the assistant hold the rod on the hub and sight backwards through the telescope. Have the assistant move a pencil up and down on the rod scale. The numbers on the rod cannot be read and the cross hairs are not visible, but it can easily be determined when the pencil point is centered in the view. Record this elevation as a backsight.

3. Rotate the scope and take a reading on the other hub 200 ft away in the normal manner. Record the elevation as a frontsight and subtract from the backsight elevation obtained in step 2. Do this several times and record the average reading.

4. Compare the difference in the elevations recorded in step 1 with those recorded in step 3. Any differences over two or three thousandths of a foot should be considered significant.

5. Adjustments to the instrument level vial must be made to correct this problem. This is a simple procedure. The field engineer should consult

the manufacturer's manual for information pertaining to adjusting individual instruments, or the instrument should be returned to the manufacturer or an authorized instrument repair shop for calibration.

Reciprocal Leveling

If it is not possible to set the instrument at a point approximately halfway between the hubs, as has been suggested, the reciprocal leveling method may be used to reduce alignment, curvature, and collimation errors. Observe Figure 2-22 while following the steps below.

1. Set up the instrument at IP#1. Take backsight and frontsight readings on TP#1 and TP#2. Record the differential elevation.
2. Set up the instrument at IP#2. Note that IP#2 is about the same distance from TP#2 as it was from TP#1. Take backsight and frontsight readings on TP#1 and TP#2 as before. Record the differential elevation.
3. Average the elevations recorded in 1 and 2 above. Record this as the true differential elevation between TP#1 and TP#2.

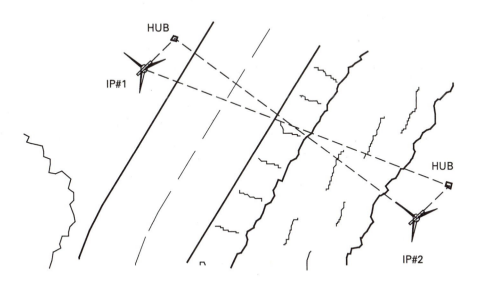

RECIPROCAL LEVELING

Figure 2-22. Reciprocal leveling.

✔ *Final Check List for Leveling Operations*

NOTE

Always check the equipment **before** leaving the storage area. Below is a short list of items which must be checked each time equipment is removed from storage for use on a job site.

1. Remove the instrument from its case and check the telescope cross hairs, bubble tubes, and all moving parts. Check the instrument case to see if all the accessories, such as plumb bobs and adjustment tools, are in place. On laser levels, check to see that the battery pack is fully charged and that the receivers are operating properly.

2. Check the tripod for proper leg extension operation. Are there any loose fittings or cracks visible? Is the tripod stable when erected? Is the instrument mounting area clean and free of nicks or scratches?

3. Assemble the instrument and tripod. Do the threads match and screw together easily? A field engineer looks foolish and wastes time and money if he or she arrives at a job site only to discover that the mounting threads on the tripod and those on the level do not match.

4. Check the level rod for proper extension movement and operation of the rod accessories. Is the rod straight? Are the markings clearly visible? Is a rod level or a rod tripod needed?

5. Will any additional equipment be required? Do you need tripod leg extensions or a tent-like apparatus to provide shade for the instrument? Will you need measuring tapes, flags, stakes, markers, field engineer's tool box, and so forth?

6. Do you fully understand the job to be done? Do you know the level of accuracy expected? Have you researched the job for the possibility of safety hazards?

If any damage to the equipment or a missing item is noted, it is the responsibility of the field engineer to inform the proper officials. *Do not attempt to use damaged equipment and do not substitute accessories from other instruments. Costly and dangerous mistakes may result.*

Chapter 2 Review Questions and Exercises

1. Define:
   a. "MSL"
   b. datum plane
   c. differential leveling

2. List at least three items which should be considered before beginning leveling operations.

3. Most instrument telescopes have two adjustments on them. What does each control?

4. List two factors that dictate the accuracy of bubble tube levels.

5. What is one major advantage of using a laser level?

6. List several major factors that must be considered in selecting and setting up a tripod.

7. What is the name of the most frequently used level rod?

8. When is "inverted rod" used?

9. When determining the difference in elevation between two points, it is recommended that the instrument be set up as nearly as possible midway between the two points. Why is this a recommended procedure?

10. When leveling an instrument, the bubble in the vial will follow the _____ thumb.

11. State the formula for differential leveling.

12. List some of the criteria for establishing the location for a job-site benchmark.

3

USING TRANSITS AND THEODOLITES

3-1   Introduction

Most leveling instruments (as discussed in Chapter 2) can only be used in the horizontal plane and do not have the capability of measuring angles. To work in both the horizontal and vertical planes and measure angles in both planes, the field engineer must use a transit or theodolite.

Unlike levels, whose telescopes will only rotate in the horizontal plane, the telescopes of transits and theodolites are free to tilt or rotate in the vertical plane as well. This feature is necessary when measuring vertical angles or when transferring construction control points or control lines to the various floors of a building under construction. Control lines and control points are discussed in depth in Part II of this text.

Transits and theodolites also have compass-like circular scales built into the instrument which allow the field engineer to accurately read angles in both planes. Most instruments have compasses built into them or are equipped with mounting points where either circular or tube compasses may be attached.

3-2   Differences Between Transits and Theodolites

Transits and theodolites perform essentially the same functions, therefore there is some disagreement as to the correct definition and description of each instrument. Some define transits and theodolites by how long the instrument design has been in use. Transits have been around for decades, while theodolites have only been in common use since the late 1950s.

Theodolites are considered by most field engineers to be an instrument of more recent design which is (1) equipped with an optical plummet, and (2) has a built-in microscopic direct-reading port or a digital display for reading angles. Angles on finer theodolites may be read to within 1 second of arc.

Transits are considered to be more traditional instruments and have fewer automatic features than the theodolite. Angles on the transit must be read directly from the circle scale. Using the vernier scales, angle reading to within 20 seconds of arc is possible with some modern transits.

Figure 3-1 shows a four-screw transit with a traditional design, a more modern three-screw transit, and a theodolite. Note that the three-screw transit and the theodolite are almost identical in appearance.

Transits are currently being manufactured that incorporate many of the features of theodolites, such as digital read-outs. As instrument technology and manufacturing techniques advance, both theodolites and transits will have similar capabilities and features. Both will have digital angle read-outs and optical plummets. The only differences will be in the precision of the instruments, principally in their angle-measuring capability.

Though modern transits and theodolites may have digital read-outs and other refinements which make using them quite easy, there are thousands of older instruments still in use. It is not uncommon to find transits well over fifty years old in daily use on construction sites. The field engineer must know how to set up and read the older instruments as well as the newer ones. Therefore, this chapter will be based on setting up and reading the scales of the older instruments. The inexperienced field engineer should be encouraged to know that once the use of the older instruments is mastered, using the newer instruments will seem relatively simple.

3-3   Equipment Setup

NOTE

Since the setup procedure for both transits and theodolites is essentially identical, the terms "transit" and "theodolite" may be used interchangeably in discussing the procedure. However, for purposes of clarity, only the term "transit" will be used in this text.

Also, the procedures for setting up the instrument on a line and over a point are almost identical. Therefore, only the slightly more difficult procedure for setting up over a point will be used as an example of equipment setup.

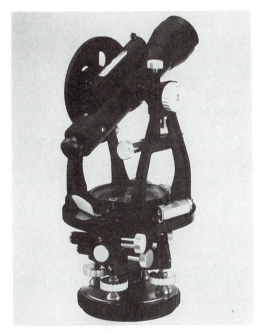

(a)

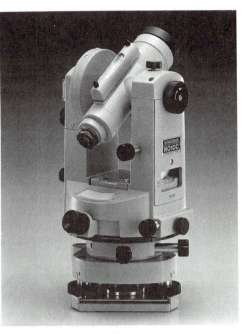

(b)

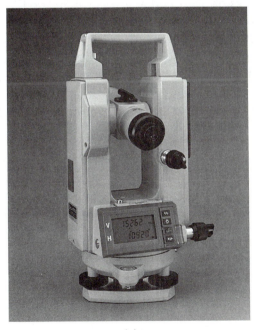

(c)

Figure 3-1. (a) Four-screw transit, (b) three-screw transit, (c) three-screw theodolite with digital read-out. (Photo courtesy of Sokkia Corporation.)

Setting up the Tripod

Tripod setup for a transit is similar to setting up for a level except that the transit is usually set up over a specific point or on a specific line. This requires that some extra steps be taken and more thought given to the actual setup of the tripod. (Review Section 2-6 as necessary.)

The two main requirements of tripod setup in locating a transit over a point are (1) that the tripod be closely centered over the point and (2) that the tripod be as level as possible. The following step-by-step procedure may be used in setting up a tripod over a point prior to mounting the transit. Some of the steps may be eliminated as the field engineer becomes more experienced in this procedure.

STEP BY STEP APPROACH

Step 1. Locate the point over which the transit is to be set up. Next, using a tape, stick, or string, roughly scribe a circle with an 18-inch radius in the dirt or on the floor around the setup point.

Step 2. Assuming the area around the setup point is reasonably level, extend the tripod legs equally to establish a comfortable working height for the instrument. Spread the legs of the tripod and set each one on the circle you scribed in step 1. Set the legs an equal distance apart around the circle. Do not push them into the ground at this time.

NOTE

If a tape is to be stretched from or across the setup point under the tripod, position the legs of the tripod on the circle so that they will not interfere with the tape.

Step 3. Using a bull's eye level or a small bubble vial level, check the transit mounting base to determine whether it is level. Check for level in two directions 90 degrees apart. If a level is not available, sight across the mounting base to a nearby level structure or sight on the horizon. Sight from one side of the tripod, then move 90 degrees to the left or right and sight again. Adjust the tripod until it is as level as possible by rotating individual legs slightly around the circle, adjusting the length of the legs, or by moving the legs in or out from the center of the circle. The type of tripod being used, the terrain, and your experience will dictate the best method to use in a specific situation.

Step 4. Remember, the transit mount area not only needs to be as level as possible, but it must also be closely centered over the setup point. A trick to use in determining whether the tripod is roughly centered over the setup point is to carefully drop several small stones, one at a time, down through the center of the transit mounting area and observe where they land. If they land consistently to one side of the setup point, the tripod must be adjusted laterally accordingly. For example, if the stones fall two inches to the left of the setup point, move each tripod leg two inches to the right. Of course, a plumb bob may be used to accomplish the same thing, but the rock trick is quicker and accurate enough for preliminary centering. The tripod should be repeatedly rough leveled and adjusted laterally until it is within about one inch of being centered over the setup point.

Step 5. Once the tripod is reasonably level and closely centered, proceed with mounting the instrument.

Setting up the Instrument

The procedure for mounting a transit on the tripod and leveling the bases of both three- and four-screw transits is identical to mounting and leveling either dumpy or wye levels. The reader is referred to Section 2-6 for a review of this procedure. Of course, the telescope of a transit has its own separate level vial mounted on it because it is free to move in the vertical plane. Therefore, the telescope as well as the base of the transit must be leveled if the transit is to be used as a level or if the horizontal plane is to be used in any way as a reference.

The rule of thumb in leveling a transit is "Start at the bottom and work your way up." This means selecting a level setup spot (if possible), level the tripod, level the base of the instrument, and then level the telescope.

Once the transit is mounted and leveled, it must be centered over the setup point. The following procedures may be used in setting up the instruments over points.

STEP BY STEP APPROACH TO USE WITH OLDER INSTRUMENTS

Step 1. Attach the plumb bob to the transit. On most older transits, a small chain or wire loop called a "bail" will be found suspended from the bottom of the instrument base in the center of the mounting area. On some instruments, a cover must be removed to expose the bail or chain. Once the bail or chain has been located, suspend a plumb bob from the bail or chain and allow it to swing freely under the instrument. (On some instruments, the plumb bob must be attached before the instrument is mounted on the tripod.) Mount the transit and level it. Adjust the height of the plumb bob until it just clears the top of the setup point. The proper method of attaching a plumb bob to facilitate height adjustment is shown in Figure 3-2.

Step 2. Use your hand to gently slow the swinging of the plumb bob until it hangs stationary under the transit. It may be necessary to fasten a sheet of plastic, scrap cardboard or plywood, or an old sheet or blanket to the tripod legs with duct tape as a temporary wind break. Observe the lateral position of the point of the plumb bob relative to the setup point. If the tip of the plumb bob is within an inch (laterally) of the center of the setup point, it will probably be possible to slide the transit on the tripod until the

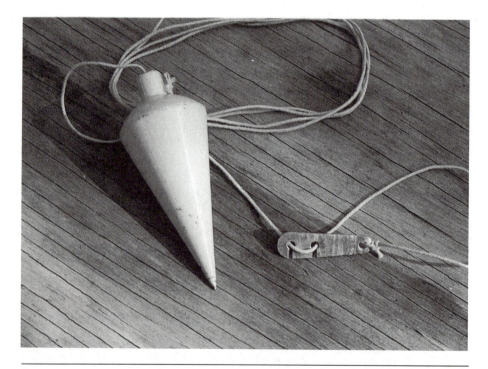

Figure 3-2.

plumb bob is centered. Gently loosen the transit mounting screw(s) or the leveling screws and slide the transit in the direction indicated by the position of the plumb bob. *Do not rotate the transit on the mount. Move it in straight lines only. Rotating the transit will throw it out of level and the leveling and centering procedure will have to be repeated from step 1.*

Step 3. Once the plumb bob is centered over the setup point, check the transit for level. Repeat steps 2 and 3 until the transit is level and the plumb bob is centered over the setup point. Tighten the mounting screw(s).

Step 4. If the transit cannot be moved far enough on the tripod for the plumb bob to be centered over the setup point, the tripod must be moved. Move the transit to the center of the mount, level it, and observe how far and in which direction the plumb bob needs to go to be centered on the setup point. Gently lift each tripod leg individually and move it in the indicated direction the indicated distance. Re-level the instrument and re-check the position of the plumb bob. Repeat steps 2 and 3 as required.

Step 5. Once the instrument is centered and level, push the tripod legs into the ground until the tripod is stable. Re-level and re-center the instrument as necessary. If the legs are pushed into the ground an equal distance, the final leveling and centering needed will be minimal. If one leg goes into the ground farther than the others, an adjustment of the length of the leg may bring the instrument to level.

Newer transits are equipped with a device called an optical plummet, which eliminates the need to use a plumb bob when setting up over a point. Using the optical plummet saves a great deal of setup time, especially in windy conditions.

Optical plummets are no more than a set of lenses and a mirror which enable the field engineer to look into a viewing port on the side of the instrument and see a point on the ground directly under the instrument. Figure 3-3 shows an instrument with an optical plummet port. The optics and mirror are installed in the lower part of the instrument so that when the base of the instrument is perfectly level, the cross hairs of the optical plummet will fall on a point precisely under the center of the instrument.

The optical plummet does the same thing that a plumb bob does on older instruments. However, there are some minor differences in the procedure used to set up these instruments. The following procedures describe the steps in setting the optical plummet-equipped instruments over a point.

Figure 3-3.   Optical plummet view port.

STEP BY STEP APPROACH TO USE WITH NEWER INSTRUMENTS

Step 1. Assuming the tripod is roughly leveled and is within about an inch of being centered over the setup point, mount the transit on the tripod but do not fully tighten the mounting screw(s). Level the transit base in accordance with Section 2-6.

Step 2. After the instrument base is level, look into the optical plummet viewing port. Focus the cross hairs by rotating the eyepiece. Some finer transits have an object-focusing adjustment which must be used to bring the setup point into focus. Some transits have no object-focusing capability. If the transit has no object-focusing adjustment, its optics are designed to focus clearly, with a 2X or 3X level of magnification, on objects within about 50 inches of the bottom of the transit.

Step 3. Observe the point where the cross hairs fall relative to the position of the setup point. **Note:** Some transit optical plummets produce an inverted image. If this is the case, you will have the tendency to move the transit in the opposite direction from which it needs to go. To prevent this, put your foot under the tripod so that you can see the toe of your shoe through the optical plummet viewing port. Compare the position of your shoe as seen in the view port with its actual position. Also, observe the point on which the cross hairs are focused in relation to the toe of your shoe. With practice, this technique will give you a clearer picture as to the position of the true center of the instrument relative to the location of the setup point.

Step 4. If the optical plummet cross hairs fall within an inch of the center of the setup point, it will probably be possible to slide the transit on the tripod until the cross hairs are centered. With the mounting screw(s) loosened, gently slide the transit in the direction indicated by the position of the cross hairs. *Do not rotate the transit on the mount. Move it in straight lines only. Rotating the transit will throw it out of level and the leveling and centering procedure will have to be repeated from step 1.*

Step 5. Once the cross hairs are centered over the setup point, check the transit for level. Repeat steps 1 through 4 until the transit is level and the cross hairs are centered on the setup point. Tighten the mounting screw(s).

Step 6. If the transit cannot be moved far enough on the tripod for the plummet cross hairs to be centered on the setup point, the tripod must be moved. Center the transit on the mount, level, and check the position of the cross hairs relative to the position of the setup point. Gently lift each tripod leg individually and move it in the indicated direction the indicated distance. Re-level the instrument and re-check the position of the cross hairs. Repeat steps 1 through 4 as required.

Step 7. Once the instrument is centered and level, push the tripod legs into the ground until the tripod is stable. Re-level and re-center the instrument as necessary. If the legs are pushed into the ground an equal distance, the final leveling and centering needed will be minimal.

3-4   Angle Measurement

Transits and theodolites may be used to measure or establish both horizontal and vertical angles. Some direct reading or digital read-out theodolites can measure both horizontal and vertical angles to an accuracy of one second of arc. Using vernier scales, many modern transits have the capability of measuring horizontal angles to twenty seconds and vertical angles to one minute of arc. Older transits, many of which fall into the antique category, can measure angles to one degree of arc.

Direct reading and digital read-out transits and theodolites are quite simple to use. The digital read-out is just like the display on a pocket calculator.

Direct reading instruments are somewhat more difficult to read. The read-out is observed by looking through a microscope on the instrument. Each manufacturer has one or more specially designed read-out windows which are available on their instruments. Figure 3-4 shows several different types of direct reading displays that are available. Since there are so many different read-out designs available and they are so easy to learn to use, they will not be examined in depth in this text. The field engineer is encouraged to consult the

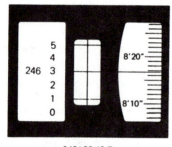

246°38′16.7″

(a)

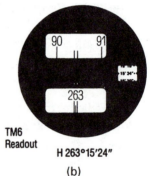

TM6
Readout

H 263°15′24″

(b)

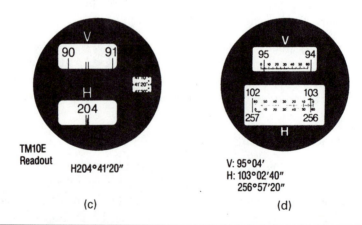

TM10E
Readout

H204°41′20″

(c)

V: 95°04′
H: 103°02′40″
256°57′20″

(d)

Figure 3-4.   Theodolite scales. (Courtesy of Sokkia Corporation.)

manufacturer's manuals if questions arise pertaining to a particular direct reading instrument.

Transits on which exposed circles and vernier scales must be used to measure angles are the hardest to learn to use. However, there are thousands of these instruments in use, so again, the field engineer must master the use of these instruments.

3-5   Reading Vernier Scales

Vernier scales are found on most transits. These scales are short, finely graduated scales which are located immediately adjacent to the horizontal and vertical circles. These vernier scales are usually accessible for viewing through a small window in the transit base. Most of the vernier division lines are so fine and so closely spaced they must be viewed through a 5X or 10X hand lens to be read with accuracy.

Figure 3-5 is a drawing of a transit circle with one minute verniers. Note that

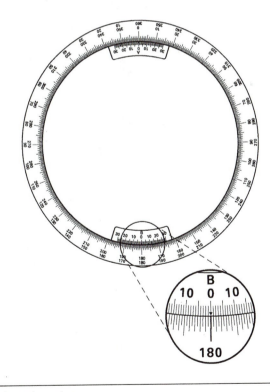

Figure 3-5.   Circle and verniers.

there are two vernier scales located 180° apart. This allows the instrument operator to use the more convenient window or to take two readings to average together for increased accuracy. The verniers are usually labeled "A" and "B."

To read vernier scales, a system of adding the readings from the main circle and the vernier is used. The rule is to work around the main circle until the index mark is reached, then jump to the vernier to complete the reading. The reading on the main circle and the reading on the vernier are added together to give the final angle measurement.

Look at Figure 3-6. This is a drawing of a "B" vernier window on a 1 minute reading transit. Follow the step-by-step procedure below to read the vernier.

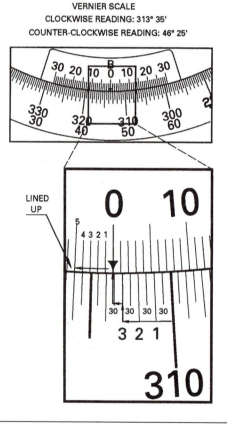

Figure 3-6.

Step 1. Decide which direction on the circle is to be read, clockwise or counterclockwise. This may be dictated by the orientation of the instrument, the direction to the backsight, or other factors.

Step 2. Assume that the circle is to be read clockwise. This means that the inner circle will be read by progressing from 300° to 310°, and so forth. Read the inner scale on the main circle from right to left (clockwise) until the vernier index pointer (arrow) is reached. Note the readings as follows:

$$310° + 3° + 30'$$

Step 3. When the vernier index pointer is reached, immediately jump to the vernier scale. Continue reading in the same direction (right to left, or clockwise, in this case), counting the marks on the vernier scale until you come to the point where a vernier scale mark and a main circle mark line up. Record this number as a minute reading and add it to the reading obtained in step 2 above.

$$310° + 3° + 30'$$
$$+ 05'$$
$$310° + 3° + 35' \text{ or } 313°35'$$

Reading the vernier scale in the counterclockwise direction, the procedure would be essentially the same. Read counterclockwise on the outer scale until the vernier index pointer is reached. Record the reading, in this case 46°, and move to the vernier scale. Proceed in the same direction until marks are found that line up. Record this point, 25', and add it to the reading from the main circle. This produces a reading of 46°25'. Note that counterclockwise and clockwise readings of verniers add to 360 degrees.

Figure 3-7 shows a circle similar to that shown in Figure 3-6. However, the circle shown in Figure 3-7 is divided into smaller segments and the vernier scales are designed to read to 30 seconds of arc. The process is the same as reading other verniers. You read the main scale until the index marker is reached, then you move to the vernier scale and read until you find two lines that are lined up.

In Figure 3-7, the process would be as follows: Assuming clockwise reading on the inside scale, the first recorded reading would be 313° plus 20'. The vernier scale reads 11' plus 30". This is added as shown below.

$$313° + 20'$$
$$11' + 30''$$
$$313° + 31' + 30'' \text{ or } 313°31'30''$$

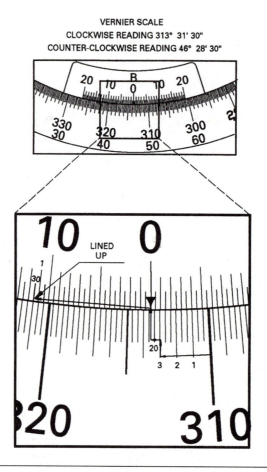

VERNIER SCALE
CLOCKWISE READING 313° 31' 30"
COUNTER-CLOCKWISE READING 46° 28' 30"

Figure 3-7.

Of course, the same process would be followed reading in the opposite direction.

There is only one way to develop speed and proficiency in reading the scales on instruments and that is to practice. One way to check on the accuracy of the reading is to read to the right, then read to the left. When the two readings are added together, the sum should be 360°. If it is not, one of the readings was taken incorrectly.

After learning to read scales, the next step is to learn to set the instrument to an angle. Some novice instrument operators find that setting an instrument to a specific angle is more difficult than reading an angle after the instrument is positioned. This procedure should be practiced, along with other angle reading procedures.

NOTE

Almost all instruments have locks which may be used to lock up the rotational movement of the various parts of the instrument. Most instruments also have fine adjustment knobs which may be used to make minute adjustments to the rotation of the body and to the elevation of the telescope of the instrument.

It is suggested that the instrument operator consult the instrument manufacturer's operation manual to learn about these adjustments. The controls will vary somewhat from instrument to instrument, but these controls are quite easy to learn to operate and their functions are essentially self-explanatory.

Chapter 3 Review Questions and Exercises

1. What is the major difference between a transit and a theodolite?

2. List three things to consider in selecting an instrument setup point.

3. Describe the procedure for locating an instrument over a point.

4. What is the major advantage of using an instrument with an optical plummet?

5. Set up two or three transits. Read the horizontal and vertical scales on each instrument. Have your instructor check your results.

6. Ask your instructor to list several directions and vertical angles at random. Using the transits set up for question 5, proceed to set the transits to the directions and angles specified by your instructor. Have your instructor check your work.

C H A P T E R

4

LINEAR MEASUREMENT

4-1 Introduction

While leveling and angle measurement provide elevation and directional control, precision linear measurement provides control of distances in both the horizontal and vertical planes. In many cases, all three—elevation, direction, and distance—must be measured with considerable accuracy to enable a construction project to progress in a timely and economic manner.

The degree of accuracy (tolerance) required in linear measurement will vary on different types of construction projects and among the various components being installed. For example, prefabricated piping or structural-steel components may have an installation tolerance of only $\frac{1}{32}$ of an inch, while a concrete highway culvert may be built several inches to one side or the other of its specified location with no serious consequences. It is the responsibility of the field engineer to determine specific job tolerance requirements by consulting the specifications and the blueprints, or by consulting the project manager.

Linear distance measurement may be conducted before, during, or after leveling and angle measurement operations. Differing degrees of accuracy may be obtained by employing different measurement methods. The most commonly used measuring methods are listed below in order of accuracy, with the least accurate listed first. Each method will be explained in detail in the following sections, with the exception of electronic distance measuring, which will be covered in Chapter 5.

1. Estimating
2. Pacing
3. Odometer
4. Stadia
5. Subtending
6. Taping
7. Electronic distance measurement (EDM)

4-2   Estimating Linear Distances

Estimating linear distances is an important skill for the field engineer to master. This skill is used daily in such cases as deciding which instruments to use, where to place control points and benchmarks, and in giving directions to others. Probably the most important use of distance estimating is as a double check against making a gross error in precision distance measurement.

As with any other skill, practice also aids the development of estimating skill. However, there are some tricks of the trade that may be employed to help the field engineer in estimating both horizontal as well as vertical distances.

Estimating Vertical Distances

There are several excellent and accurate methods used to estimate the height of objects such as buildings, trees, poles, and so forth. A quick estimate of the height of a building may be done by counting the number of stories and multiplying by 10 feet. Of course, if there is a point or spire on top of the building or if the average story height is closer to 12 feet, the estimate may be off. But the 10-feet-per-story constant may be used for very rough preliminary estimating when necessary.

If the object to be measured does not have a constantly recurring division (like the floors of a building) that can be counted, the comparison method may be used to estimate height. For example, if the height of a football-stadium bleacher is unknown, a nearby structure of known height (such as the goal posts) may be used to comparatively measure the bleacher. See Figure 4-1.

To do this, imagine standing near the end of the field facing the goal post. Hold a pencil (or similar object) vertically in the fist at arm's length and extended toward the goal post. Align the top of the pencil and the top of the goal post. At the same time, place the thumb on the surface of the pencil and adjust it until it is aligned with the bottom of the goal post. Next, extend the arm toward the stadium, aligning the thumb with the bottom of the bleacher

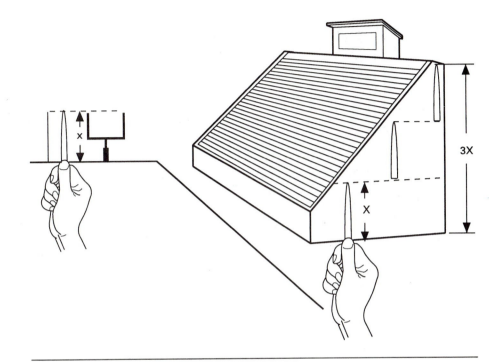

Figure 4-1.   (Illustration by Dale Barnes.)

area. Sight across the top of the pencil to a point on the bleacher. Mark this point with the eye and raise the pencil until the thumb is at the point just marked. Repeat this procedure until the top of the stadium is reached. Multiply the height of the goal post by the number of sightings it took to reach the top of the stadium. Adjust the outcome as necessary if the final sight across the pencil does not come out exactly even with the top of the stadium.

Another similar comparison method using a pencil or similar object also involves sighting across the top of the pencil and the thumb. See Figure 4-2. The arm is extended toward the structure and the top of the pencil is aligned with the top of the structure. The thumb is aligned with the bottom of the structure.

Next, the fist is rotated 90 degrees until the pencil is horizontal. The thumb is aligned with the bottom of the structure and some object is sighted across the top of the pencil. The distance between the base of the structure and the object sighted across the top of the pencil may be measured or paced for an estimated height of the structure.

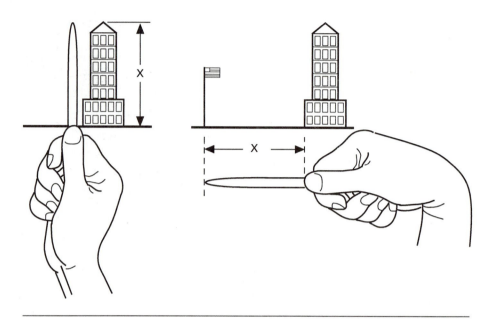

Figure 4-2. (Illustration by Dale Barnes.)

A third, but more accurate, method of estimating vertical height of a structure involves comparing the shadow of the structure with the shadow of an object of known height. Figure 4-3 shows the principles and mathematics involved.

In Figure 4-3(a), the known height of the goal post and the measured length of its shadow are compared to the measured length of the stadium shadow to determine the height of the stadium. In Figure 4-3(b), the height of a stake and the length of its shadow are used in the same manner to determine the height of the building.

NOTE

The measurements of the shadows should be taken quickly so that the length of the shadows do not change appreciably. Also, the terrain where each shadow falls must be similar for the comparison to be accurate.

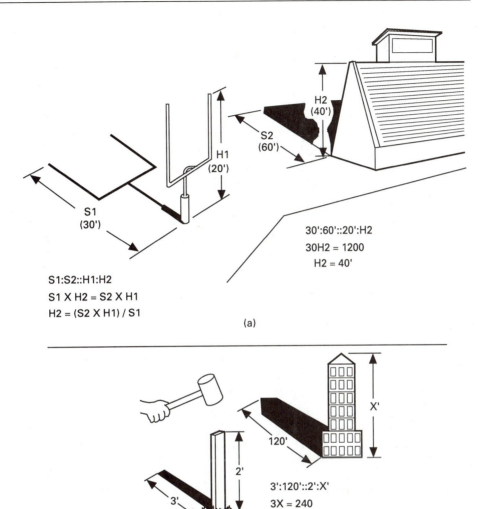

H2
(40')

S2
(60')

H1
(20')

S1
(30')

30':60'::20':H2
30H2 = 1200
H2 = 40'

S1:S2::H1:H2
S1 X H2 = S2 X H1
H2 = (S2 X H1) / S1

(a)

2'

120'

X'

3'

3':120'::2':X'
3X = 240
X = 80'

NOTE: All dimensions must be in like units.

(b)

Figure 4-3.   (Illustration by Dale Barnes.)

Estimating Inaccessible Horizontal Distances

Often, horizontal distances which are inaccessible, such as bodies of water or busy highways, must be estimated. Using a pencil extended at arm's length and other comparison methods similar to those explained above may be used if it is possible to get to a vantage point such as a hillside or a tall building near the area to be measured.

Another method is to stand at the edge of the inaccessible area and pull a hat or cap down over your eyes until the brim of the hat is aligned with the edge of the opposite side of the inaccessible area. See Figure 4-4. Then, without moving your head up or down, turn your body until an accessible point at a similar elevation can be sighted under the hat brim. Now measure the distance between where you are standing and the point just sighted. To aid in keeping your head from moving up and down during this procedure, it may be helpful to make a fist and hold it under your chin while turning from the inaccessible point to the accessible point.

4-3   Pacing

A handy skill to combine with estimating distances is pacing. It is not uncommon to find field engineers and others who can measure horizontal distances by pacing and be accurate to within two or three feet in a hundred yards. For many, it is a matter of pride to be able to measure quite accurately by pacing.

To measure by pacing, the field engineer must first know the length of his or her average step. The following procedure may be used to determine this.

1. Set up a pacing course. A vacant lot or parking area can be used. A straight, unobstructed area of at least 100 feet is recommended. The course should be level and should not be in loose dirt or sand. An unlined athletic practice field is ideal.
2. Stretch a tape alongside the course. The person whose pace is to be measured should stand at the starting point, focus on an object at or beyond the end of the course, and start walking toward that object. The tape should not be watched.
3. As the person walks, the steps taken must be counted. As the end of the course is approached, the person walking should count aloud as each step is taken. The tape should not be watched and care must be taken not to modify the pace to finish exactly at a specific point. A second person should observe the last few paces, mark the exact end of one specific step near the end of the tape, and note the count at that point.

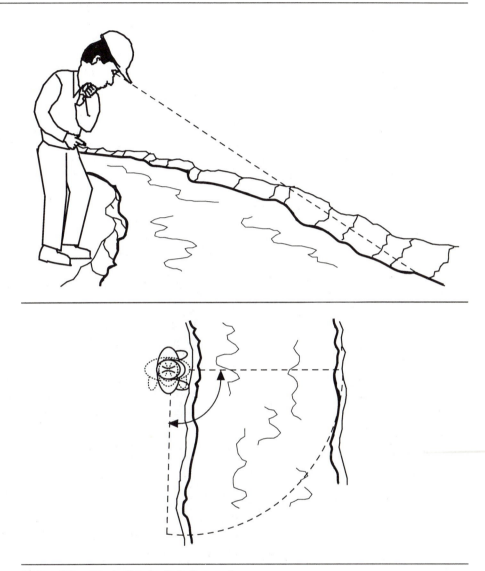

Figure 4-4.   (Illustration by Dale Barnes.)

The person pacing should continue on for another step or two to ensure an evenly paced course.

⊕ **NOTE**

A normal pace should be maintained. The steps should not be extended or shortened from what would be a normal "walking with a purpose" pace. This means walking as if going to a specific place rather than taking an evening stroll in the park. Also, construction working shoes should be worn, not clogs or heels!

---

4. After measuring from the starting point to the point marked in step 3 above, divide the distance covered by the number of steps taken to find the length of each step. For example, if the pacer took 40 paces and the observer noted 100 feet at the end of the 40th step, the pace length would be 2.5 feet or 30 inches.

5. Repeat the exercise several times and average the results. Memorize the results. Use some association method to remember this figure. For example, if the length of the step averaged out to be 28 inches, this might be associated with someone's birthday on 2-28 (February 28).

6. Some people can successfully modify their pace to obtain accurate results. For example, if the average pace is 34 inches, the pacing could be consciously stretched so that the pace equals 36 inches or 3 feet. This makes calculating distances by the pacing method much easier, but the technique must be practiced.

4-4 Odometers

The most familiar odometer is that found on automobile speedometers. The smallest graduation on the standard automobile odometer is one tenth of a mile or 528 feet. Attempting to estimate readings between the graduations on the instrument is quite difficult, so using an automobile to measure short linear distances is not very accurate.

However, special digital odometers for automobiles are available which are accurate to within one foot in one mile. These accessory odometers are relatively inexpensive and may be easily moved from one vehicle to another. A simple calibration procedure is performed to maintain accuracy if the odometer is moved from one vehicle to another or if new tires or tires of a different diameter are installed. Some of these accessory odometers will display distances in feet, miles, meters, or kilometers. The more-expensive odometers may be hooked up to recording devices from which data can be down-loaded into compatible office computers. This enables the computer to prepare reports reflecting the data gathered by the odometer.

Hand-operated odometers are available for use when measurement must be done outside vehicles. These odometers are no more than a wheel attached to a handle. (See Figure 4-5.) Most are equipped with a counter which registers the revolutions of the wheel. The operator sets the counter to 0, and then walks along the path to be measured, pushing the wheel ahead like a small cart. The counter reading is taken at the end of the path. Of course, muddy or uneven ground conditions will affect the reading, but accuracy to within a few inches in 100 yards can be obtained under good conditions.

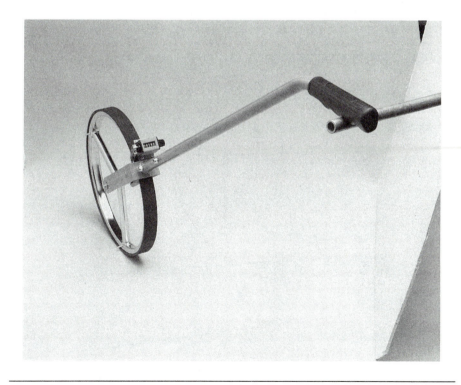

Figure 4-5.   Odometer wheel. (Photo courtesy of Sokkia Corporation.)

4-5 Stadia Measurement Techniques: Horizontal and Inclined

Optical measurement of linear distances can be accomplished by using an instrument equipped with stadia hairs. The stadia hairs are two short horizontal hairs located one above and one below the center cross hair. (See Figure 4-6.) Some older instruments and inexpensive instruments manufactured recently do not have stadia hairs, but most modern levels and transits are so equipped.

Stadia distance measurements can be quickly and easily performed with an accuracy to within a few inches in one hundred yards. Stadia measurement is often used in preliminary surveying or in topographic mapping when more precise measurements are not needed. The stadia procedure may also be used to double check for gross errors when taping or when using other more precise measuring methods.

Horizontal Stadia Measurement

The stadia hairs are positioned on the vertical cross hair so that when a rod is sighted through the telescope, the difference in elevations observed at the top

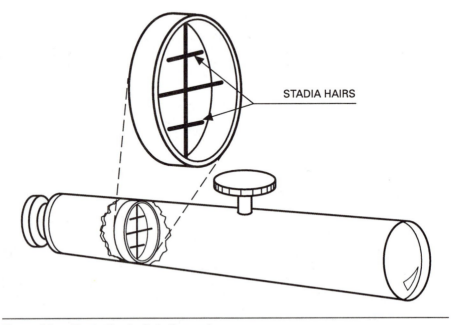

STADIA HAIRS

Figure 4-6. (Illustration by Dale Barnes.)

and bottom stadia cross hairs is 1/100th of the distance between the instrument and the rod. (See Figure 4-7(a).) Therefore, if the difference in elevations between stadia hair observations is multiplied by 100, the result would be the distance from the instrument to the rod. The figure 1:100 is called the stadia ratio.

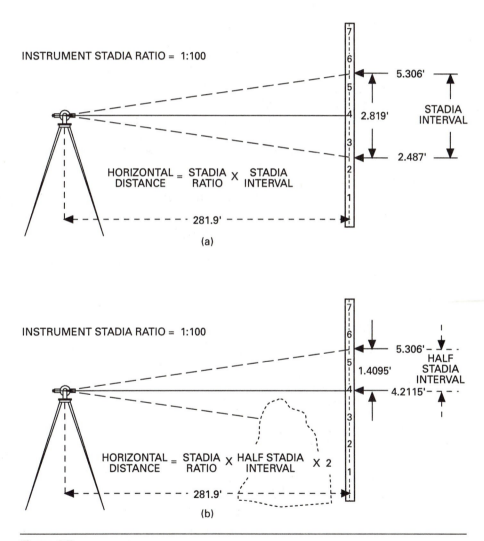

Figure 4-7.

EXAMPLE

Refer to Figure 4-7(a). With the instrument (level or transit) properly set up and leveled, a rod is sighted. The elevation observed at the top stadia hair is 5.306 ft. The elevation at the bottom stadia hair is 2.487 ft. Subtract the bottom reading from the top reading. The result, 2.819 ft, is called the stadia interval. Multiply the stadia interval (2 .819 ft) by 100 to obtain the horizontal distance of 281.9 ft.

$$5.306 \text{ ft}$$
$$-2.487 \text{ ft}$$
$$2.819 \text{ ft} \times 100 = 281.9 \text{ ft}$$

The result, 281.9 ft, is the distance from the instrument vertical center line to the rod measured at the height of the horizontal center line of the instrument telescope. Note that no adjustments for changes in elevation have been made at this point.

Using half stadia. Refer to Figure 4-7(b). In the example above, both the upper and lower stadia hairs are used in determining the distance from the instrument to the rod. However, in many cases, the view of the rod may be obscured by trees or bushes and both stadia readings cannot be taken. In this case, a rod reading at the center cross hair and one stadia hair can be taken. The half stadia interval is multiplied by the stadia ratio and the result is multiplied by 2.

NOTE

Most modern instrument manufacturers have standardized on the 1:100 stadia ratio as discussed above. However, there are some instruments that have a different ratio and some that may also have correction factors (called instrument constants) that must be added to the distance calculations. See Figure 4-8(a,b).

The field engineer should consult the manufacturer's manual to determine the stadia ratio and instrument constant for a particular instrument. This and other information about an instrument may also be found displayed on a printed form fastened to the inside of the instrument case.

If the stadia ratio or instrument constant (if any) are unknown, a rod should be set up 100 ft from the instrument and stadia interval readings taken.

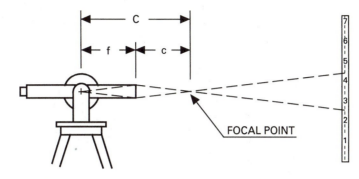

Some older instruments and instruments with long telescopes have a focal point some distance in front of the object lens. The distance from the focal point to the center of the instrument, represented by "C" above, must be added to the stadia distance calculation.

(a)

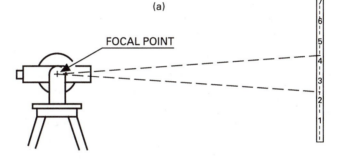

Newer short telescope instruments have a system of mirrors inside the telescope which causes the focal point to fall almost at the center of the instrument. The "C" factor for these instruments is so small that it is considered to be 0.0'. In this case, nothing is added to the stadia distance calculation.

(b)

Figure 4-8.

The observed stadia interval should be 1 ft if a 1:100 ratio instrument is being used. If the stadia interval is slightly under 1 ft, say .987 ft, the instrument focal point is probably some distance in front of the telescope. This distance, the instrument constant, must be determined if accurate stadia measurements are to be taken with this instrument.

To determine the unknown value of the instrument constant, multiply the stadia interval by the stadia ratio and subtract the result from 100 ft: (.987 ft × 100 = 98.7 ft) then (100 ft – 98.7 ft = 1.3 ft). The result, 1.3 ft in this case, is the instrument constant. This value must be added to each stadia measurement. Repeated stadia measurements of known distances should be done to prove these calculations.

Using stadia to establish lines, arcs, or circles. Horizontal linear stadia distance measurement procedures may be used to establish rough lines, arcs, and circles by simply reversing the stadia measurement procedure.

EXAMPLE 1

Suppose a circle is to be laid out with a radius of 250 ft. A level or transit is set up over the center point of the circle. The rod holder moves to a point on the proposed circle at which the instrument operator can get a 2.5 ft stadia interval reading on the rod. The rod position is marked and the rod holder moves around the circle, repeating the process until the required points have been marked.

EXAMPLE 2

Suppose the right-of-way boundaries of a road must be roughly established. With the instrument set up on the center line, the instrument operator would turn the telescope to a position 90° to the center line of the road. The rod holder would proceed to a point at which the instrument holder could read the proper stadia interval on the rod. The position would be marked. At this point, the scope could be plunged or rotated 180° and the procedure repeated to establish the opposite boundary line. This procedure can be used to roughly lay out parallel or offset lines at any time.

Inclined Stadia Measurement

Generally, anytime stadia measurements must be taken on terrain with a slope of more than three degrees, inclined stadia measurement methods are employed. If the slope is less than three degrees, the horizontal stadia measurement method is used and the slope is disregarded. Additionally, differential elevation measurements can be taken at the same time as the linear measurements with little extra effort.

To use stadia distance and elevation measurement techniques on sloping ground, mathematical calculations must be done which take into consideration the instrument height, the vertical angle between the horizontal plane and the instrument line of sight, and the elevation of the line of sight intercept point on the rod. As vertical angles must be measured when inclined stadia measurements are taken, transits are used instead of levels.

The field data collection procedures used when inclined stadia measurements are taken differ somewhat from customary field data collection techniques. The main difference is that practically no taping, pacing, or other direct linear measuring is done on the ground. This speeds up the operation considerably, but results in a corresponding loss of accuracy. However, one must remember that stadia measurement is designed for rough preliminary measurements or to check for gross errors when using other, more-precise methods of measurement.

The following is a step-by-step procedure for using a transit to gather the field data necessary to establish the distance and differential elevation between two points using the inclined stadia method. The calculations necessary to complete the field work are discussed immediately following the field procedures, as are field book entries.

Field Procedures

Step 1. Assuming the preliminary work of selecting the proper points for benchmarks, monuments, and so forth has been done, proceed to the first transit setup point. This setup point should be a permanent or temporary benchmark of known elevation, either actual or assumed. Set up and level a transit over the point selected. Note the setup point and elevation in the field book.

Step 2. On the instrument, set the upper motion and the circle to 0°. Lock the upper to the circle, leaving the lower lock free. Have the rod holder proceed to the point selected for the first rod reading. The rod holder should plumb the rod, use high rod, wave the rod, and so forth as directed by the instrument operator to obtain a good rod reading. Sight the rod and lock the lower part of the instrument.

Step 3. After focusing on the rod, the instrument operator should refer to the vertical angle scale on the transit. The vertical angle should be adjusted slightly until the nearest division on the circle scale and the vernier index pointer on the vernier scale are aligned. This minor adjustment will have no effect on the distance and elevation calculations, but it diminishes the chances of error which often occurs when reading vernier scales. It also makes finding the sines and cosines of the angles easier and reduces the need for interpolation in many cases. The angle of elevation should be recorded in the field book along with the location and elevation of the transit setup point.

NOTE

Some field engineers adjust the vertical angle of the instrument until the cross hairs rest on the rod at the same elevation as the HI elevation. This eliminates some steps in the calculations done later.

Step 4. Next, the instrument operator should take three readings from the rod; one at the top stadia hair, one at the center cross hair, and one at the bottom stadia hair. The difference between the top and bottom stadia readings is recorded as the stadia interval (SI) in the field book. The center cross hair reading is recorded as the rod reading (RR) or, as it is sometimes called, the rod intercept (RI).

NOTE

It is permissible to adjust the vertical movement of the telescope so that the top stadia hair falls on an even foot or tenth mark on the rod. This aids in the calculations and does not significantly affect the accuracy of the sighting. The rod observations may be repeated as necessary to obtain a more accurate sighting. In some cases, the average of the sightings is used in the calculations that are done later.

Step 5. Unless the height of the instrument (HI) has been established by backsighting, measure the distance from the top of the benchmark under the transit vertically to the optical center of the instrument to obtain the HI distance. Most instruments will have a mark, such as a line or a cross on the telescope barrel, that indicates the optical center. Of course, the measurement will not be precise due to having to measure around the base of the instrument, but this difference will be negligible for stadia purposes. Record the "HI" figure in the field book.

Step 6. Unlock the upper part of the instrument and rotate to sight on another benchmark or other object with a known location on or near the job site. Record in the field book which benchmark or object was sighted and the azimuth reading between the rod and the benchmark or other object in the field book.

NOTE

Step 6 may be done prior to observing the rod if so desired. This step is designed to provide a positive check on the instrument location for future references.

Step 7. Review the operation for obvious errors. Have the rod holder move to another location and repeat steps 2 through 6 as required. Note the new location in the field book. If no other stadia observations are to be made from this setup point, move to another setup point, repeating steps 1 through 6. If this completes the field data collection process, secure the equipment in accordance with the recommended procedures and proceed with the calculation process.

Calculation process (stadia reduction). The process of performing the calculations related to stadia measurement is called stadia reduction. There are two methods of performing stadia reduction calculations to derive the horizontal distances and the differential elevations required in inclined slope stadia measurements. One method uses formulas only and the other uses formulas supplemented by a set of tables called stadia reduction tables. Stadia slide rules and scales built into the instrument vertical circles have been used in the past to aid in stadia measurement calculations, but today hand-held calculators and computer programs are more often used to solve stadia formula problems. However, not all field engineers will have access to the latest in computer equipment, so both the formula and the table method of stadia reduction will be examined in this text.

STADIA REDUCTION BY CALCULATION

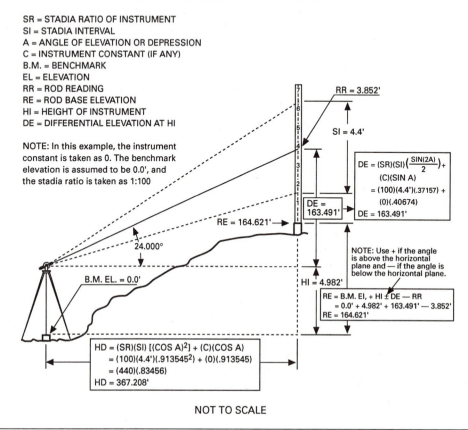

NOT TO SCALE

Figure 4-9.

Figure 4-9 shows a typical inclined stadia problem. It is obvious that if the rod could be held perpendicular to the line of sight of the instrument and if other measurements could be easily determined, this would be a simple problem. However, it is not practical or possible to hold the rod perfectly perpendicular to the line of sight. To overcome this problem, the rod is held in the normal vertical position and mathematical calculations are performed to determine the true horizontal distances and elevations.

There are three basic formulas that are used in stadia reduction calculations. These formulas are used to calculate (a) the horizontal distance (HD) from the transit to the rod, (b) the difference in elevation (DE) between the height of the instrument and the point where the center cross hair intercepts the rod, and (c) the elevation at the base of the rod (RE).

The various data used in the formulas are gathered in the field or are taken from the instrument specifications.

The various entities in the formulas are:

HD = The horizontal distance from the instrument to the rod.

DE = The difference in elevation between the height of the instrument telescope (HI) and the point where the center cross hair intercepts the rod (RR).

RE = The elevation at the base of the rod.

SR = The stadia ratio of the instrument.

SI = The stadia interval (difference between top and bottom stadia hair readings on the rod).

A = The angle of elevation or depression of the instrument telescope.

C = The instrument constant (if any).

B.M. = The elevation of the benchmark under the instrument.

EL = Elevation

RR = Rod reading at center cross hair. (Sometimes called the rod intercept.)

HI = Elevation of the line of sight through the optical center of the instrument.

The formula to calculate the horizontal distance (HD) is stated as:

$$HD = (SR)(SI)[(COS\ A)^2] + (C)(COS\ A)$$

The formula to calculate the difference in elevation (DE) between the instrument height (HI) and the rod reading (RR) at the center cross hair is stated as:

$$DE = (SR)\ (SI)\left(\frac{SIN(2A)}{2}\right) + (C)\ (SIN\ A)$$

The formula to calculate the rod elevation (RE) at the base of the rod is stated as:

$$RE = B.M.\ EL + HI \pm DE - RR$$

Note: Use + if the vertical angle is elevated. Use − if the vertical angle is depressed.

Refer to Figure 4-9. In this example, the field data and instrument specifications are:

$$SR = 100 \qquad SI = 4.4 \text{ ft}$$
$$A = 24 \text{ degrees} \qquad C = 0 \text{ ft}$$
$$B.M. = 0.0 \text{ ft elevation} \qquad RR = 3.852 \text{ ft}$$
$$HI = 4.982 \text{ ft}$$

This data inserted into the formulas above yields:

FOR HORIZONTAL DISTANCE:

$$HD = (SR)(SI)[(COS\ A)^2] + (C)(COS\ A)$$
$$= 100 \times 4.4 \text{ ft} \times .913545^2 + 0 \text{ ft}$$
$$= 440 \times .83456$$
$$HD = 367.208 \text{ ft}$$

FOR DIFFERENTIAL ELEVATION:

$$DE = (SR)\ (SI)\left(\frac{SIN(2A)}{2}\right) + (C)\ (SIN\ A)$$
$$= 100 \times 4.4 \text{ ft} \times .37157 + 0 \text{ ft}$$
$$DE = 163.491 \text{ ft}$$

FOR ROD ELEVATION:

$$RE = B.M.\ EL + HI \pm DE - RR$$
$$= 0.0 \text{ ft} + 4.982 \text{ ft} + 163.491 \text{ ft} - 3.852 \text{ ft}$$
$$RE = 164.621 \text{ ft}$$

Refer to Figure 4-9 again. This figure presents all the necessary formulas for stadia reduction and for differential elevation calculations. It also presents an example problem using the data above, with the results noted on the profile drawing.

Stadia table method. The use of stadia tables simplifies the stadia reduction process by simplifying the formulas. The tables are used in place of the cumbersome method of manipulating the sines, cosines, and stadia ratios, as is done in the calculation method. The rod elevation formula remains the same.

Refer to the stadia tables in Appendix C. Entities used in the stadia reduction formulas by the table method are:

VALUE A = From the horizontal distance column for angle A in the stadia reduction tables.

VALUE B = From the differential elevation column for angle A in the stadia reduction tables.

VALUE C = From the "C" values at the bottom of the appropriate stadia reduction table.

The formula to calculate the horizontal distance (HD) is stated as:

$$HD = (SI)(VALUE\ A) + (VALUE\ C)$$

The formula to calculate the difference in elevation between the instrument height and the rod reading at the center cross hair is stated as:

$$DE = (SI)(VALUE\ B) + (VALUE\ C)$$

The formula to calculate the elevation of the rod at the rod base (RE) is the same as before:

$$RE = B.M.\ EL + HI \pm DE - RR$$

Remember to use + if A is elevated and − if A is depressed.

Refer to Figure 4-10. The same field data and instrument specifications as previously used inserted into these formulas yields the following:

FOR HORIZONTAL DISTANCE:

$$HD = (SI)(VALUE\ A) + (VALUE\ C)$$
$$= 4.4\ ft \times 83.46 + 0\ ft$$
$$HD = 367.224\ ft$$

FOR DIFFERENTIAL ELEVATION:

$$DE = (SI)(VALUE\ B) + (VALUE\ C)$$
$$= 4.4\ ft \times 37.16 + 0\ ft$$
$$DE = 163.504\ ft$$

Refer to Figure 4-10 again. This figure presents all the necessary formulas for stadia reduction and for differential elevation calculations using the stadia tables. It also presents an example using the data above, with the results noted on the profile drawing. Note that there are some minor differences in the resulting linear measurement calculations and elevation calculations. This is the result of rounding at different stages in the formulas and will have no appreciable effect on the job.

STADIA REDUCTION BY TABLES

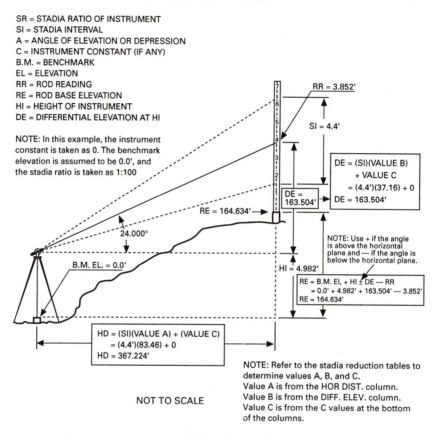

SR = STADIA RATIO OF INSTRUMENT
SI = STADIA INTERVAL
A = ANGLE OF ELEVATION OR DEPRESSION
C = INSTRUMENT CONSTANT (IF ANY)
B.M. = BENCHMARK
EL = ELEVATION
RR = ROD READING
RE = ROD BASE ELEVATION
HI = HEIGHT OF INSTRUMENT
DE = DIFFERENTIAL ELEVATION AT HI

NOTE: In this example, the instrument
constant is taken as 0. The benchmark
elevation is assumed to be 0.0', and
the stadia ratio is taken as 1:100

RR = 3.852'

SI = 4.4'

DE = (SI)(VALUE B)
    + VALUE C
    = (4.4')(37.16) + 0
DE = 163.504'

DE =
163.504'

RE = 164.634'

24.000°

NOTE: Use + if the angle
is above the horizontal
plane and — if the angle is
below the horizontal plane.

B.M. EL. = 0.0'

HI = 4.982'

RE = B.M. El, + HI ± DE — RR
   = 0.0' + 4.982' + 163.504' — 3.852'
RE = 164.634'

HD = (SI)(VALUE A) + (VALUE C)
   = (4.4')(83.46) + 0
HD = 367.224'

NOTE: Refer to the stadia reduction tables to
determine values A, B, and C.
Value A is from the HOR DIST. column.
Value B is from the DIFF. ELEV. column.
Value C is from the C values at the bottom
of the columns.

NOT TO SCALE

Figure 4-10.

Stadia Reduction on the Computer

There are many computer programs available for surveying. Most of these programs will have a stadia reduction component that will allow the entry of field data, perform the calculations, and print the results. If the field engineer has access to such a program, he or she should become familiar with the program prior to gathering the field data. The field engineer can then modify the field data gathering and recording process to facilitate data entry into the computer.

Appendix D of this text contains a simple computer program for stadia reduction. It is written in the GWBASIC language. Should the student or field engineer wish to use this program, it may be typed into the user's computer. Use care to type it exactly as it is written. Some minor translation may be needed to adapt to other forms of BASIC, so due caution must be used. An experienced programmer may easily modify or expand this program to suit his or her needs.

CAUTION

After entering the program, thorough testing using a proven set of data is mandatory before using the program on a stadia reduction project.

Stadia Field Notes

As with any field measurement operations, accurate field notes of the stadia observations must be recorded. Figure 4-11 shows the field notes of a stadia

Stadia Survey - Reynolds Property
Lots 5, 6, 7, - Industrial Park   Page 1 of 2
⊼ @ B.M. EL. = 100.00' H.I. = 4.982'

BS to 1	Hor <	S.I.	Vert <	Hor. Dist.	Diff. Elev.
Rod @ 1	0°0'0"	1.483'	5°6'	147.1'	13.13
Rod @ 2	66°21'	.935'	-1°20'	93.5'	-2.17
Rod @ 3	109°36'	1.749'	-3°27'	174.3'	-10.5
Rod @ 4	151°10'	1.789'	5°13'	177.4'	16.20
Rod @ 5	177°54'	2.594	12°46'	246.7'	55.91
Rod @ 6	200°42'	2.164'	13°29'	204.6'	49.06
Rod @ 7	205°31'	3.093'	17°55'	280.0'	90.54
Rod @ 8	207°58'	3.858'	22°45'	328.1'	137.59
Rod @ 9	219°5'	3.807'	22°19'	325.8'	133.7
Rod @ 10	230°13'	3.921'	22°07'	336.5'	136.76
Rod @ 11	233°15'	3.165'	14°43'	296.1'	77.77
Rod @ 12	240°14'	2.268'	14°55'	211.8'	56.42
Rod @ 13	270°3'	1.123'	6°54'	110.7'	13.39
Rod @ 14	188°0'	.837'	6°21'	82.7'	9.21
Rod @ 15	220°2'	2.036'	14°22'	191.1'	48.94

⊼ R. Sward K&E #6   page 2 of 2
C. Hargrove Rod #4   3-13-92
R. Sward   cold-appx. 40 deg
                                    cloudy, calm

RR	RE
1.632	116.48'
2.981	99.83'
3.746	90.73'
5.190	115.99'
4.909	155.98'
3.916	150.13'
12.564	182.96'
8.341	234.23'
7.973	230.73'
5.823	235.92'
3.398	179.35'
4.127	157.27'
5.275	113.10'
4.789	109.39'
4.880	149.04'

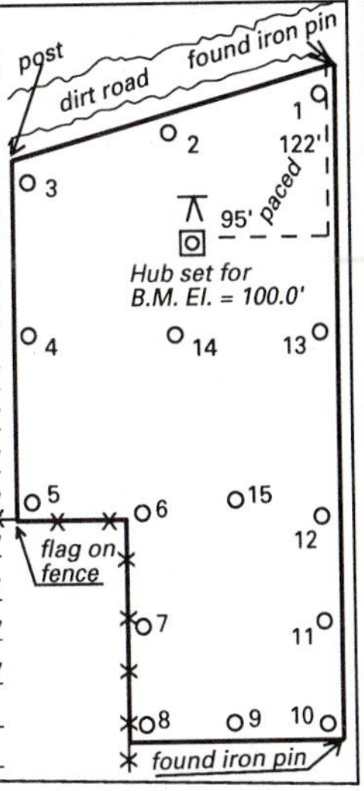

post
dirt road
found iron pin
○1 122'
○2
○3 95' paced
⊼
◻ Hub set for
B.M. El. = 100.0'
○4 ○14 13○
○5 ○6 ○15 ○12
flag on fence
○7 11○
○8 ○9 10○
found iron pin

NOTE: The Hor. Dist., Diff. Elev., and RE column entries may be calculated and entered in the office after the field data have been entered.

Figure 4-11.

survey in which elevation, angle, and distance readings along the boundaries and at selected locations in the center of a piece of property were taken.

Note the general information at the top of the pages and the drawing on the right. Note also the headings of the columns. All data were entered in the field except the Hor. Dist., Diff. Elev., and the RE column data. These data were calculated by formula, stadia table, or computer program.

Stadia Cautions

As with any field work, attention to detail will prevent having to return to the field to repeat readings. Here are some points which will help you to perform error-free work.

1. Do not confuse the stadia hair with the center cross hair. Write down the three readings in order. Observe the sequence. The center cross hair reading should be approximately halfway between the stadia hair readings. If not, read the rod again.
2. Read the rod twice and the angles twice. Look at them carefully. Do the readings appear reasonable or has a gross error been made? Is there an appreciable difference between the first readings and the second readings?
3. Double check your math. Do the stadia interval readings appear reasonable for the distance involved? Compare them to other similar distances.
4. It is recommended that stadia sightings be kept to distances of less than 300 feet.

4-6   Subtending

Subtending as a means of linear measurement is similar in principle to stadia measurement, in that measured angles and mathematics are used to determine distances between points. However, there are two major differences in the equipment and the process. First, the measurement of the angle is done in the horizontal plane only. Secondly, the target on which the instrument is sighted is a bar of a known fixed length called a subtense bar.

In practice, an instrument point and a rod point are selected using criteria similar to selecting points for stadia measurement. A tripod is erected over the point where a rod would normally be set. A subtense bar is mounted on the tripod and leveled with a built-in bubble level. The bar is then rotated to a position which is perpendicular to the line of sight of the instrument. To aid in aligning the bar perpendicular to the instrument line of sight, most subtense bars will have a small telescope or other sighting device built into the bar at its center.

The subtense bar has two targets mounted permanently on it, one at either end. The centers of these targets are precisely two meters apart. Temperature compensation devices in the bar keep the distance between the targets as constant as possible.

The subtense bar is sighted after leveling and being brought to a position perpendicular to the instrument line of sight. The horizontal angle (A) between the targets is measured. The horizontal distance (D) is calculated using the formula:

$$D = \frac{1}{2}S \times COT \frac{A}{2}$$

(S is the distance between the subtense bar targets.)

Figure 4-12 shows a typical subtending problem using a subtense bar.

Some subtense bar manufacturers furnish a set of tables with their bars that give the linear distances (D) associated with the various horizontal angles (A). These tables are similar to the stadia tables that were discussed earlier and

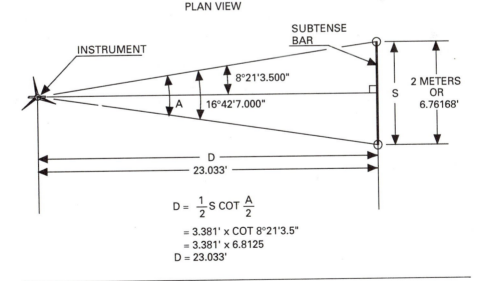

Figure 4-12.

are available in either meters or feet. To use the tables, the bar targets are sighted and the horizontal angle between the targets is measured. The angle is looked up in the tables and the corresponding distance read. This eliminates the need to do any calculation using logarithms and formulas.

Accurate linear measurement using the subtense bar requires an instrument capable of measuring horizontal angles to one second of arc. Such instruments are quite expensive. Economically priced EDM equipment capable of linear measurement to similar or greater levels of accuracy is rapidly replacing the subtense bar.

Even though the subtense bar is not being used as regularly as in the past, the principles of subtending may be used to roughly determine linear distances using a slightly different technique.

Figure 4-13 is the plan view of a typical subtense problem except that a known distance "S" is used in the formula rather than the standard 2-meter distance between the subtense bar targets. In this example, the formula is worked out in the space below the plan view and all the field data and calculated values are shown on the plan view.

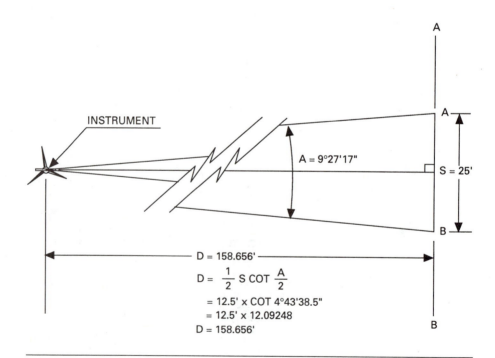

$$D = \frac{1}{2} S \cot \frac{A}{2}$$

$$= 12.5' \times \cot 4°43'38.5''$$

$$= 12.5' \times 12.09248$$

$$D = 158.656'$$

Figure 4-13.

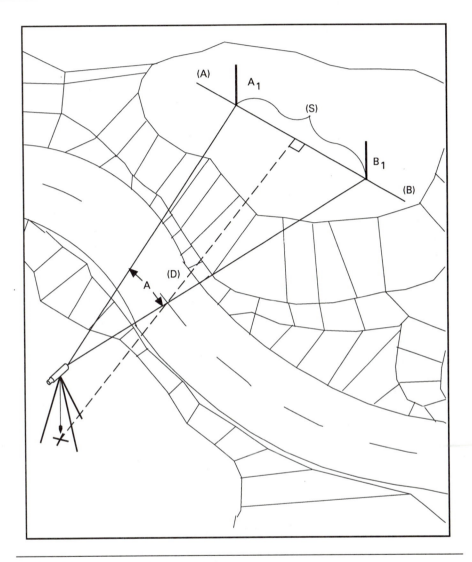

Figure 4-14. (Illustration by Dale Barnes.)

Figure 4-14 shows another approach to a subtense problem. In this example, points A1 and B1 are established on line AB by using any precise linear measuring technique, usually by taping. The midpoint of line A1B1 is found and marked. The 90-degree angle is turned with the instrument as it sits over the midpoint of line A1B1. Point X is found and staked.

Next, the instrument is moved and set up over point X. Points A1 and B1 are sighted and angle A is measured. The distance D from point X to the midpoint of the line between points A1 and B1 is calculated using the same formula as used in subtending.

$$D = \frac{1}{2}S \times \cot \frac{A}{2}$$

4-7   Taping (Chaining)

Precision linear measurement using tapes is one of the first skills learned by the surveyor. This process was originally called "chaining" because a metal chain was used to measure distances. (See Section 1-6.) Some surveyors still refer to linear distance measurement as "chaining," even when modern tapes or EDM equipment are used.

Though it sounds relatively simple, using a tape to determine linear distances to a high degree of accuracy is somewhat complicated and requires study and practice to develop proficiency. Taping levels of accuracy are expressed as a ratio such as 1:3000 for construction work or 1:100,000 for high-precision geodetic and scientific surveys. These ratios are simply specifications for the upper allowable limits of error in a specific situation.

EXAMPLE

The specifications call for the location of a building on a job site to be accurate to a ratio of 1:3000. The plans show that the building should be 183 ft from a property line. It is actually built 183.042 ft from the line. Is it located within the acceptable limits?

The error is .042 ft (about 1/2 in.) in 183 ft. Expressed as a ratio, this is .042:183. How does this compare to the allowed 1:3000? The ratio of .042 ft to 183 ft must be converted to a 1:X form for comparison purposes. The mathematics required to do this conversion are shown below:

$$.042:183::1:X$$
$$.042X = 183$$
$$X = 183 \div .042$$
$$X = 4357$$
$$.042:183 \text{ is equivalent to } 1:4357$$

These calculations show the building to be placed within an accuracy of 1:4400 (rounded to the nearest 100 from 4357), which is well above the 1:3000 limit. However, had the building specification called for a placement accuracy of 1:5000, the location of the building would not have been within limits.

To put the specifications into perspective, a specification of 1:50,000 would have required the building to be less than ³⁄₆₄" out of position.

The use of precision taping techniques has become more common on construction projects in recent years. This is due to the more-demanding aspects of design and the use of prefabricated materials and components that must fit perfectly into buildings and other structures.

Choosing a Tape

The variety of tapes available from manufacturers today is quite extensive. Tapes made of cloth, fiberglass, and steel are available, as are tapes made of a composite of materials. Tapes from 10 ft to 500 ft in length are commonly available, but the tapes most used in construction work are 100 ft and 300 ft in length. Tapes with graduations in meters, feet and inches, or feet and tenths of a foot are available. The design and spacing of the graduations may be selected when ordering tapes. Special orders are taken by some manufacturers.

Most tapes are furnished with a reel on which the tape is wound for storage. However, some heavy-duty tapes are furnished without reels. These tapes are carefully coiled and stored in metal or plastic cases.

Choosing the right tape for a specific job depends on the accuracy requirements as well as the distances and type of terrain to be covered. More exacting measurements (those in the 1:50,000 accuracy category, for example) will call for precision tapes and the use of precision taping techniques. Most construction measurement work will be in the 1:3000 accuracy range and a lesser degree of precision in the taping techniques will suffice.

One of the first decisions in selecting a tape is to decide on the length needed. As previously stated, most construction work can easily be done with a 100-foot or 300-foot tape. Most of these tapes are supplied wound on a reel which may be in the form of an open framework or an enclosed case. (Some field engineers prefer the open reels because they allow the tape to dry out faster if has been used in wet grass or other damp conditions.) The tape itself may be easily disconnected from the reel assembly for replacement when worn or broken.

Once the decision has been made about how long the tape should be, the

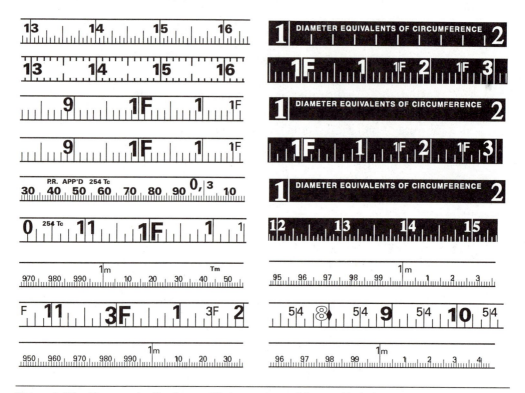

Figure 4-15. Tape graduation types. (Photo courtesy of Cooper Tools.)

type and style of graduation should be considered. Figure 4-15 shows some of the tape graduation styles available. In the United States, most construction project drawings show some dimensions in feet and inches and others in feet and fractions of a foot expressed as a decimal. The dimensions of the structure itself are usually shown in feet and inches. Land measurements, ground elevations, and foundation component elevations are shown in feet and fractions of a foot. Most field engineers will have at least two tapes available, one graduated in feet and inches, and one graduated in feet and tenths. The tape that more closely corresponds with the dimensions on the plans should be used.

Most manufacturers have tapes available with both feet and inches and feet and tenths of a foot graduations on the same tape. While these tapes with dual graduations may be convenient in some cases and might seem to eliminate the need to maintain two different tapes, there is the possibility that readings might be inadvertently confused. Extreme caution must be used not to confuse the readings if, indeed, tapes with dual graduations are used.

⊕ **NOTE**

Some chief field engineers and project managers discourage the use of the dual-marked tapes and in some instances will not allow them on the job site.

Some tapes have rather large graduations in the center portion of the tape and much finer graduations near either one or both ends, usually in the first and/or last foot. For example, some 100-foot tapes will only be marked every foot in the area between one foot and ninety-nine feet. However, the first and last foot will be graduated in tenths and perhaps hundredths of a foot. These tapes are sometimes called "cut" tapes. See Figure 4-16(a).

To perform accurate measurement, an even foot mark on the body of the tape is aligned with one layout point while the second layout point is aligned with one of the fine graduations within the first foot. The distance between the layout points is derived by reading the even foot mark at the first layout point and deducting the distance from 0 on the tape to the second layout point. Cut tapes are becoming less prevalent due to the chance of error in the subtraction process.

On some tapes, the fine graduations are located in a one-foot long area before the 0 foot mark. These are called "add" tapes. See Figure 4-16(b). As in using a cut tape, the layout points are aligned with an even foot mark on the tape and a point next to one of the fine graduations. The distance between the layout points is derived by reading the even foot mark on the first layout point and adding the distance from 0 on the tape to the second layout point.

Tape manufacturers also have tapes available that have fine graduations along their entire length. While these tapes are more expensive, experience has shown that the added expense is usually justified because these tapes speed up the operation and reduce errors.

Along with the length and type of graduation markings, the quality and level of precision of the tape should be considered. The width of the graduations, the accuracy of their placement, and the thermal properties of the tape contribute to the tape's precision.

The graduation markings on tapes with lower precision are quite wide, while the graduation markings on high-precision tapes may be no wider than .003". The high-precision tapes are manufactured and the graduations placed on the tape in more closely controlled processes. These tapes are usually calibrated with a master tape maintained by the manufacturer or are calibrated by comparing them with a standard tape maintained by some state and federal agencies.

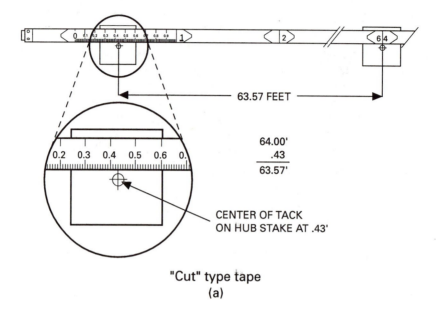

63.57 FEET

$$\begin{array}{r} 64.00' \\ \underline{.43} \\ 63.57' \end{array}$$

CENTER OF TACK
ON HUB STAKE AT .43'

"Cut" type tape
(a)

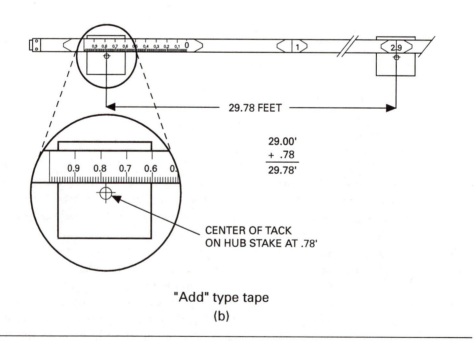

29.78 FEET

$$\begin{array}{r} 29.00' \\ + \underline{.78} \\ 29.78' \end{array}$$

CENTER OF TACK
ON HUB STAKE AT .78'

"Add" type tape
(b)

Figure 4-16.

One of the main problems encountered in precision taping is the contraction and expansion of the tape with changes in temperature. The length of an ordinary 100-foot steel tape may change .01 foot with a 15°F change in temperature. High-precision tapes are made of alloys which change very little with temperature fluctuations, perhaps as much as 1/40th of the amount of an ordinary steel tape. These high-precision tapes are often purchased by surveyors and contractors and used for nothing else than to check and calibrate other less-expensive tapes. Invar, lovar, and minvar are names associated with some high-precision tapes.

Taping Accessories

Tensioners, clamps, thongs, thermometers, taping pins, repair kits, and other taping accessories are available from tape manufacturers. Each of these items is designed to make the taping process easier and more accurate. Figure 4-17 shows several tape accessories. Most of these accessories are available from

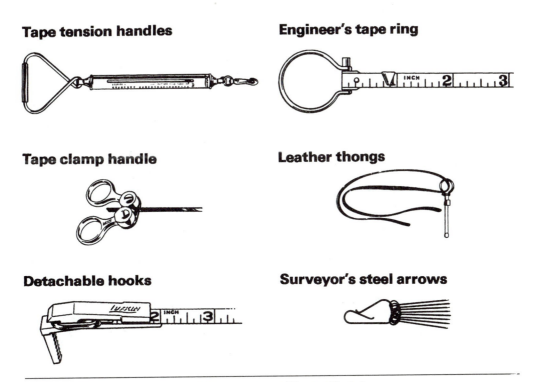

Tape tension handles

Engineer's tape ring

Tape clamp handle

Leather thongs

Detachable hooks

Surveyor's steel arrows

Figure 4-17.   Tape accessories. (Photo courtesy of Cooper Tools.)

surveyors' supply stores, construction equipment suppliers, or engineering and architectural supply stores.

Tensioners. Most precision tapes are designed to be used under a certain amount of tension. All materials will stretch a certain amount when put in tension, and tapes are no exception. To make sure that the amount the tape stretches is as close as possible to being the same each time the tape is used, tensioners are used. With the tensioner, which is in reality a small scale, the field engineer can put the same amount of tension on the tape every time it is used, therefore ensuring equal stretch with each use. This ensures an equal amount of stretch with each use to put a given amount of pull on the tape.

Clamps and thongs. Since most tapes are thin metal ribbons, they are quite hard to grip when stretching them over rough terrain. Clamps are used to grip the tape anywhere along its length without the danger of cutting one's hand or bending and breaking the tape. Leather thongs are used in the end rings of tapes to allow easier gripping in that region.

Thermometers. Thermometers designed to be attached to the tape are often used to take tape temperature readings when it is necessary to correct for thermal expansion or contraction of the tape.

Taping pins. Taping pins are metal rods about ³⁄₁₆" in diameter and about 14" long and come in sets of 11. (This is enough to mark every 100 feet on a 1000-foot line.) The rods are sharpened on one end and have a large round ring formed in the other. They are usually painted with alternating red and white stripes. Sometimes called surveyor's arrows or chaining pins, these rods are used to mark temporary positions during the taping process.

Repair kits. Repair kits are available that allow tapes to be spliced when broken. These kits are supplied with materials and instructions that can be used for accurate splicing. In some cases, tapes must be sent back to the manufacturer for repair and re-calibration.

Manufacturers' services. Tape manufacturers often provide specialized custom services for their customers. Special length tapes with special graduation markings can be ordered. The manufacturers can calibrate tapes under special conditions of support, sag, and temperature. Most will furnish certificates of accuracy with all their products.

Precision Taping Techniques

The techniques used to obtain accuracy in the 1:50,000 range are quite extensive and are seldom necessary in construction work. An accuracy in the 1:3000 range is normally considered to be quite sufficient for most construction work. However, to ensure that the 1:3000 range is easily reached, many field engineers use techniques designed to produce a 1:5000 of level accuracy. The time necessary to hold to the more stringent specifications is negligible and the results are well worth the effort.

Below is a list of six basic requirements for taping with an accuracy of 1:5000. These requirements apply to every 100 linear feet measured.

1. Use a calibrated tape. Accuracy should be within .01 ft in 100 ft of true length with the amount of deviation known. Manufacturers furnish a data sheet with each precision tape they sell. This data sheet will tell precisely how accurate the tape is and to which standard it is compared. Today, even moderately priced 100-foot tapes are usually accurate to within .01 ft.

2. Tape temperature during measurement must be known to within plus or minus 5°. Manufacturers usually calibrate their tapes with the temperature of the tape at 68°F. In the field, it is not possible to maintain this ideal temperature, so mathematical corrections must be made to allow for tape expansion and contraction.

Most steel tapes have an expansion coefficient of 0.0000065 ft per foot per degree Fahrenheit. This is equivalent to 0.00065 ft on a 100-foot tape with each single degree of temperature variation. To compensate for temperature-induced changes in tape length, the total distance measured (in feet) is multiplied by the coefficient of expansion (0.0000065). The result is then multiplied by the algebraic difference between the observed temperature minus the manufacturer's calibration temperature.

NOTE

Take the tape temperature at the tape. The temperature of a tape lying on a hot asphalt road or on cool wet ground may be considerably different from the temperature of the surrounding air.

EXAMPLE

A tape calibrated at 68°F is used to measure the distance between two points. The tape reading was 236.89 ft. A tape thermometer showed a tape temperature of 87°F during taping. What is the actual distance measured?

$$C_t \text{ (temp. correction)} = 236.89 \text{ ft} \times 0.0000065 \times (87°F - 68°F)$$
$$= .00154 \times 19$$
$$C_t = .029 \text{ ft}$$

The C_t value of .029 ft would be added to the 236.89 ft to give a distance corrected for tape temperature of 236.92 ft (rounded to the nearest hundred). Therefore the distance measured would be recorded as 236.92 ft, even though the tape reading was 236.89 ft. These figures should be shown in the field book. Tape temperatures below 68° require that the C_t value be subtracted from the measured distance.

A layout problem would be worked out in almost the same way. The only difference is that the C_t value would be subtracted from the measured distance when tape temperatures are over 68°F and added when tape temperatures are below 68°F.

EXAMPLE

A distance of 376.45 ft is to be laid out. The tape temperature is 79°. What distance should be measured with the tape to yield an actual distance of 376.45 ft?

$$C_t \text{ (temp. correction)} = 376.45 \text{ ft} \times 0.0000065 \times (79°F - 68°F)$$
$$C_t = .027 \text{ ft}$$

The C_t value of .027 ft would be subtracted from 376.45 ft. A taped distance of 376.42 ft would be marked. Again, these figures should be recorded in the field book.

3. The manufacturer's suggested tension must be applied to the tape during measuring. Tension must be within plus or minus 2 pounds of what the manufacturer suggests.

Manufacturers calibrate their tapes under a certain amount of tension with the tape supported along its entire length. If the tape is used in the field and is supported at the ends only, the natural sag in the middle of the tape will cause the actual distance to be shorter than the distance read from the tape. There are several ways to adjust for the sag in the tape.

One way sag corrections are made is by increasing the tension on the tape. This increases the stretch of the tape and decreases the sag to a point where it is negligible. This practice is usually sufficient for construction surveying work. Most tape manufacturers furnish tables with each tape which give the tape correction figures with the tape supported, not supported, and with various tension forces applied.

A second way to correct for tape sag is by using the following mathematical formula:

$$C_s \text{ (Sag length correction)} = -\frac{W^2 L}{24 P^2}$$

where

W = weight of the tape (not including reel or case)

L = unsupported length of the tape

P = the amount of tension applied

EXAMPLE

A distance of 367.35 ft is measured in stages with a 100-ft tape supported only at its ends. The tape weighs 5 pounds and the manufacturer's suggested tension of 20 pounds is applied. What is the sag correction factor?

The first step is to calculate the sag in each 100-ft-long run of the tape. This is done by plugging the values from the example into the formula.

$$
\begin{aligned}
C_s &= -\frac{W^2 L}{24 P^2} \\
&= -(5^2 \times 100) \div (24 \times 20^2) \\
&= -(25 \times 100) \div (24 \times 400) \\
&= -(2500 \div 9600) \\
C_s &= -.260 \text{ ft per 100 ft of tape}
\end{aligned}
$$

As there were three 100-ft measurements taken, multiply the value derived above by 3.

$$-.260 \text{ ft} \times 3 = -.780 \text{ ft sag in the 300-ft area}$$

Next, calculate the sag for the 67.35-ft long area. Use only 67.35% of the weight of the tape (3.3675 pounds) and the length of 67.35 ft in this calculation. The P value (20 pounds) remains the same.

$$C_s = -\frac{W^2L}{24P^2}$$
$$= -(3.3675^2 \times 67.35) \div (24 \times 20^2)$$
$$= -(11.340 \times 67.35) \div (24 \times 400)$$
$$= -(763.749 \div 9600)$$
$$C_s = -.080 \text{ ft sag in the } 67.35\text{-ft area}$$

The last step is to add the −.780 ft value and the −.080 ft value.

$$-.780 \text{ ft} + -.080 \text{ ft} = -.86 \text{ ft}$$

The .86 ft would be subtracted from the 367.35 ft to give a distance corrected for tape sag of 366.49 ft.

A layout problem would be approached in a similar manner, except that the sag correction factor is added to each tape run.

EXAMPLE

A distance of 366.49 ft is to be laid out using a 5-pound tape and 20 pounds of tension. Use the same formula as in the example above.

Three 100-ft lengths would be laid out by measuring 100.260 ft in each length. Using the remaining 66.49 ft as the L value in the formula, the C_s value for the last 66.49 ft is calculated to be .078 ft. Adding the .078 ft to the 66.49 ft gives 66.57 ft. The final measurement would be 66.57 ft.

NOTE

The algebraic sum of all corrections (C_t, C_s, and so forth) to tape readings is applied to the observed measurement to determine the actual distance.

4. The tape must be read when in the horizontal position or, if read on a slope, slope correction calculations must be used to determine the true horizontal distance. Both methods are described below.

Taking horizontal tape readings on an incline. To take horizontal tape readings on a slope, a technique called "breaking tape" is used. In this procedure, the tape readings are taken in "steps" down the incline. In most cases, only a relatively short portion of the tape is used in each step.

⊕ **NOTE**

Some field engineers prefer to take the head of the tape down the incline while others prefer to reverse this procedure. The head-downhill method is described below, but the reverse of this procedure may be used when necessary.

The head tape person moves down slope only as far as he or she can comfortably hold the tape in a horizontal position and still apply the correct tension. This usually puts the head end of the tape about chest high. Figure 4-18(a) shows the details of this procedure.

When the tape is horizontal and tensioned, a plumb bob is lowered from the head end of the tape until it is about two inches above the ground. The plumb bob string is draped over the tape, preferably at the "0" point, and held in place with the thumb. See Figure 4-18(b). The tape may be dipped slightly to stop the swing of the plumb bob.

When the tape is horizontal and tensioned and both the head tape person and the rear tape person agree that a reading should be taken, the head tape person releases the pressure on the plumb bob string allowing the plumb bob to drop to the ground. The tape reading at the point where the plumb bob string was held on the tape is recorded as the head tape measurement reading. As noted earlier, this would ideally be the "0" point on the tape. The point where the point of the plumb bob struck the ground is marked with a taping pin as the point from where the rear tape person will begin the next step in the taping process.

When the tape is at an awkward height for easy tensioning, a range pole may be used as a lever to aid in applying a constant tension to the tape. Figure 4-18(c) illustrates this procedure.

Taking tape readings up or down inclines. Tape readings are taken in the usual manner, with tension and temperature corrections being calculated. However, corrections for the degree of the slope must be made as well. To make corrections for taping on a slope, either the angle of the slope or the difference in elevation between the bottom and the top of the slope must be known. In some instances, the difference in elevation at the end of each whole or partial tape length is needed. Differential elevations are determined to the nearest .01 ft by leveling techniques. Angles are determined by transit observations. See Section 2-7 for information on leveling techniques and Section 3-4 for information on angle measurement using transits.

(a)

(b)

(c)

Figure 4-18. (a) Breaking tape, (b) ready to drop plumb bob, (c) using a range pole and thong to hold tension.

Method 1—Slope correction based on a known angle of elevation.
    The horizontal distance can be calculated by the formula:

$$D_h \text{ (horizontal distance)} = D_s \cos A$$

where

    D_s = distance measured on the slope
    A = angle of elevation

EXAMPLE

A 128.43 ft distance is measured on a 6° slope. What is the true horizontal distance?

$$D_h = D_s \cos A$$
$$= 128.43 \text{ ft} \times .99452$$
$$D_h = 127.73 \text{ ft}$$

The horizontal distance corrected for slope is 127.73 ft.

Method 2—Slope correction based on a known differential elevation.
    The horizontal distance can be calculated by the formula:

$$D_h \text{ (horizontal distance)} = \sqrt{(D_s^2 - D_e^2)}$$

where

    D_s = distance measured on the slope
    D_e = difference in elevation between bottom and top of slope

---

EXAMPLE

A 128.43 ft distance is measured on a slope. The difference in elevation between the benchmark at the top of the slope and the one at the bottom of the slope is 13.43 ft. What is the true horizontal distance?

$$D_h = \sqrt{(D_s^2 - D_e^2)}$$
$$= \sqrt{(16494.2649 - 180.3649)}$$
$$D_h = 127.73 \text{ ft}$$

5. Alignment along the traverse should be maintained by transit rather than by sighting on distant range poles. If several tape positions are to be observed on a particular traverse, a transit should be set up over the first point and sighted on the last point. Intermediate points should be kept in alignment by sighting with the transit.

6. All readings should be taken and recorded to the nearest .01 ft. Care must be taken in rounding and in estimating tape readings. Most field engineers carry trigonometry functions and other calculations to five decimal places, but round the resulting distances to the nearest .01 ft.

STEP BY STEP APPROACH

The following is a step-by-step method of precision taping for attaining 1:5000 or better accuracy.

Step 1. Check the necessary equipment before going to the field. This includes tripod, transit, level rod, hand level, plumb bobs (2), taping pins (set), tape (decimal feet, or feet and inches, or both), tape accessories (tensioner, clamps, thongs), field notebook, felt markers, spray paint, stakes, hammers, flags or ribbons, communication equipment, and any necessary safety equipment.

 NOTE

A taping party is usually made up of two or three people: the head tape person, the rear tape person, and a recorder. In the case of two-person parties, the head tape person usually acts as the recorder. Any member of the party can be considered the party "chief," which is the title given to the person in charge of a surveying party.

Step 2. The party chief should assure that the objective of the taping procedure is clear and that all relative information is available to the party. This includes information about the site, the location of benchmarks, property lines, corners, and so forth.

Step 3. Upon arrival at the job site, the party chief should check in with the project manager or others in charge at the location for a briefing on possible conflicting or hazardous activities that might be in progress. This check-in procedure also enables the project manager to warn others who may come later of the presence of the taping party. After checking in, the party chief should explore the area and the route to be used for the taping procedure. Benchmarks, property lines, and corners should be located and marked as necessary. Any hazards such as traffic, construction equipment in operation, and other potential dangers should be eliminated or marked. The members of the party should be briefed on safety concerns, communications, individual responsibilities, and the general objectives of the operation.

Step 4. Once the taping route has been identified, the transit should be set up on the line to be taped. Three possible transit locations can be used: (1) some point between the starting and ending points, (2) a few feet in front of the starting point, or (3) a few feet beyond the ending point. Any of these positions will work as long as the entire tape route is visible from the transit position. If erected between the starting and ending points, the tripod legs must be arranged so as not to interfere with the taping procedure.

Step 5. The head tape person gives the rear tape person one taping pin, takes ten taping pins and the "0" end of the tape, and heads toward the far end of the line to be taped. A tape thermometer is attached to the tape and allowed to stabilize. Alignment of the head tape person's position is directed by the instrument operator.

Step 6. Once the tape is on line, the head tape person applies proper tension and the rear tape person aligns the 100-ft mark with the starting point. When the rear tape person has the 100-ft mark precisely on the starting point, a signal is given to the head tape person. The head tape person then marks the 0-ft point with a taping pin or other marking device. If a taping pin is used, the pin is slanted slightly in a direction away from the tape. If nails, keel, paint, or felt-tip pens are used for marking, a taping pin is left pointing to the marked point.

Step 7. After the proper notes are made in the field book (see Figure 4-19), the rear tape person picks up the taping pin and moves toward the first marked point. The head tape person moves toward the second point to be marked. Alignment is again directed by the instrument operator.

Step 8. Steps 6 and 7 are repeated until the line has been taped from beginning to end. After the last point is marked, the head and rear tape persons count the taping pins that have been picked up by the rear tape person. The number of pins should equal the number of times the tape length was read during the operation (assuming a pin is not used at the final point). To double check the figures and to reduce the possibility of a gross error, the taped line may be paced, hurriedly re-taped, or measured by stadia observations.

Step 9. Elevation readings or slope angle measurements should be done as necessary and entered in the field notebook.

Step 10. Review the operation and field notes for obvious errors. If this concludes the field operations, secure the equipment in accordance with the recommended procedures. The party chief should notify the project manager that the taping operation has been completed. The party chief should then present the results of the assignment and file the field notes as required.

Taping Cautions

As with any field work, attention to detail will improve accuracy and prevent costly errors.

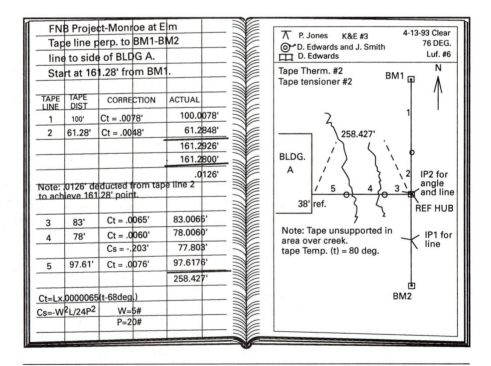

Figure 4-19.

1. If a tape is used on which the "0" point falls at the end of a loop or other fitting on the end of the tape, determine exactly where the true "0" point is and use it. Some field engineers, in an attempt to improve accuracy, will "burn" a foot on these types of tape. That is, they will use the 1-ft mark as the 0-ft point. A 100-ft tape is then effectively reduced to 99 ft in length and each point on the tape is 1 ft short of its true position. This practice is discouraged because more errors occur than the supposedly improved accuracy warrants.

2. Tape downhill if possible. It is easier for the head tape person to properly tension the tape.

3. Position yourself on the side of the tape where the numbers can be read in their normal position. It is quite easy to incorrectly read a number on a tape if it is viewed upside down.

4. When reading the tape, start at a clearly marked point a division or two before the point to be measured. Call out the numbers as your eyes approach the final division. This natural reading of numbers from smaller to larger will aid in preventing the transposition of numbers.

5. Tape twice whenever time permits.

6. Do not drag a tape across a power cord. The edge of the tape may cut through the insulation of the cord.

7. Do not tape across rail lines. Some are electrically charged. Also, some will register an obstruction on the tracks if a tape causes a direct short between the two rails. This may disrupt rail traffic for many miles in both directions and may cause an accident.

8. Perform taping operations early in the morning if possible, before the sun has a chance to expand building components. This practice also will aid in negating the effect of heat expansion of the tape. If this is not possible, try to do repeated measurements of the same distances at the same time each day. Try to stay either entirely in the shade or entirely in the sun when taping a line. This makes the temperature of the tape more uniform and makes calculations for tape expansion more accurate.

Students are encouraged to obtain catalogs from tape and surveying equipment manufacturers. A thorough study of these catalogs to see what is available in the way of tapes and accessories is highly recommended. Often a special tape or an accessory item can be purchased that will save a great deal of time and money. It is the field engineer's responsibility to know what is available on the market in order to plan and purchase proper equipment with which to do special jobs effectively.

Chapter 4 Review Questions and Exercises

1. Demonstrate to your instructor two ways to estimate linear distances and two ways to estimate vertical distances.

2. What are stadia hairs?

3. What is the stadia ratio?

4. What is the stadia interval?

5. A stadia interval of 2.872 ft is observed through an instrument which has a stadia ratio of 1:100. What is the distance from the instrument to the rod? Assume no incline.

6. Refer to the formulas in the text and Figures 4-9 and 4-10 to determine the horizontal distances (HD), differential elevation (DE), and the rod elevation (RE) for each of the following problems. Use the stadia formula method.

(a)	SR = 100	(b)	SR = 100	(c)	SR = 100
	A = 28°		A = 13.56°		A = 9.25°
	B.M. = 0.0' elev.		B.M. = 197.0' elev.		B.M. = 245.5'
	HI = 4.5'		HI = 3.90'		HI = 4.876'
	SI = 3.0'		SI = 4.75'		SI = 2.91'
	C = 0'		C = 0'		C = 0'
	RR = 3.0'		RR = 4.32'		RR = 3.567'

7. Refer to the data in problem 6 above. Calculate the HD, DE, and RE values using the stadia table method.

8. Determine the distance (D) from an instrument to a point using the subtending method where s = 40 ft and A = 36.15033°.

9. A distance of 255.79 ft is to be laid out using a tape whose temperature in the field is known to be 75°F. What distance would you measure with this tape to achieve the actual distance of 255.79 ft assuming that the tape is supported along its entire length?

10. A certain tape weighing 4.5 pounds is designed to read 100.0 ft with 20 pounds of tension applied when supported at the ends only with a tape temperature of 68°F. Considering only the sag of the tape, what would the amount of the adjustment be to ensure that an actual distance of 255.79 ft is laid out?

11. Why is it recommended that taping be done in the early morning or late afternoon?

5

ELECTRONIC DISTANCE AND ANGLE MEASUREMENT

5-1   Introduction

The use of electronic surveying equipment has become commonplace on today's construction job site. However, in its early stages of development, electronic surveying equipment was so expensive that only large engineering and surveying companies and government agencies could justify purchasing such a system. However, in recent years, this equipment has moderated in price so that most construction companies can now readily afford it. The electronic distance measuring instrument (EDMI) can be put into use quickly and can produce more accurate results than is possible with conventional surveying equipment. Some systems can provide an electronic record of the work done. With interfacing hardware and software, the EDMI can be coupled with computer systems to provide print-outs and plots of field work.

Perhaps the easiest way to understand the operation of the EDMI is to relate it to the use of sonar and radar. The sonar, radar, and EDMI operate on the principle of either sound or light waves being transmitted, bounced off an object, and the return wave being compared to the transmitted wave. The internal physics of the operation of EDMIs is of secondary importance to the field engineer. However, he or she must realize that sound and light waves can be obstructed, deflected, and modified by weather and adverse job-site conditions. These conditions must be recognized and accounted for when using EDMIs.

The beams transmitted by EDMIs must be accurately reflected back to the instrument for proper reading and comparison. This is accomplished by using a reflecting prism. A mirror could be used, but perfect alignment of the transmitter and the reflecting mirror would have to be maintained. This would be quite difficult if not impossible to do in the field, so a prism assembly is used to reflect the light beams back to the instrument. Some misalignment can be tolerated using a prism, without affecting accuracy. Figure 5-1 shows both a single-prism and a multiple-prism assembly. The multiple-prism assembly is used in long-range distance measurement while the single prism is used in shorter-range applications.

Most EDMIs have built-in calibration devices that automatically adjust the instrument to account for changing conditions. However, some of the earlier instruments and some of the less-expensive instruments produced today require the operator to calibrate the instrument with each use. This usually involves simply entering data such as temperature, wind conditions, and so forth into the internal computer of the instrument. This is done by punching in the required information on a data entry pad mounted on the instrument. This data entry pad resembles the pad on a hand calculator. See Figure 5-2.

The various stages of development of computers and other electronic equipment are referred to as "generations." As previously stated, the early generations of EDMIs were naturally quite large, crude, inaccurate, and required a great deal of power to operate. Second-generation EDMIs were transistor-equipped, smaller, and required less power. Third-generation EDMIs used micro-chips, were highly mobile, and could be operated on the power supplied by small battery packs.

Each generation of EDMIs took several forms. Some were add-on accessory items designed to be used in conjunction with conventional transits and theodolites. See Figure 5-3. Others were totally self-contained units capable of angle measurement in both vertical and horizontal axes as well as conventional and electronic distance measurement. See Figure 5-2.

The current generation of EDMIs consists of many of the bolt-on accessory types as well as self-contained units. Most of today's EDMIs are capable of producing an electronic record of field work. This electronic record, usually in the form of a magnetic card no larger than a credit card, can be inserted into

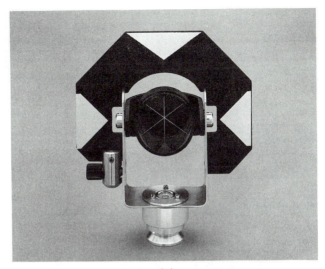

(a)

(b)

Figure 5-1. Single- (top) and multiple- (bottom) prism assemblies. The multiple-prism assembly is mounted on a tribrach. (Photo courtesy of Sokkia Corporation.)

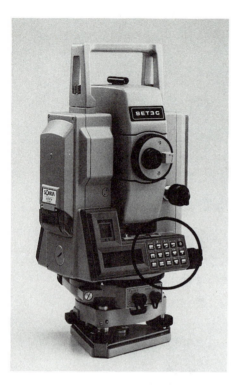

Figure 5-2. Electronic distance measuring instrument. Circle indicates the data entry pad. (Photo courtesy of Sokkia Corporation.)

a reading device which is connected to a desk-top computer. The field data can then be used to generate reports or to produce plots of the property surveyed.

The term "total station" is used to refer to instruments that have all the features described above.

The distance measuring accuracy of an EDMI is specified by manufacturers by their use of a term similar to: "+ or − (3mm+3ppm)." This term means that the accuracy is 3 mm + 3 parts per million of the distance being measured. For example, if a line one million millimeters long were to be laid out, the distance actually measured with the EDMI should be within + or − 6 millimeters of the true length of 1,000,000 millimeters.

Today, EDMIs specially designed for construction job sites are being produced by many manufacturers. The cost of these instruments is comparable to mid-range theodolites. Many of these EDMIs are capable of measuring angles to within .5" of arc and linear distances to within ¼" in 3,000 ft or greater. The superior accuracy, ease and speed of operation, and the moderating prices

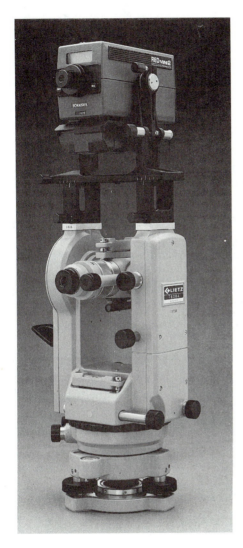

Figure 5-3. Add-on electronic distance measuring instrument mounted on a theodolite. (Photo courtesy of Sokkia Corporation.)

make these instruments ideal for construction work. Their use is spreading rapidly.

Using EDMIs, the construction field engineer can perform all the common tasks associated with site development, construction surveying, and building layout in record time. In some cases where job-site visibility is ideal, an entire project can be laid out from one instrument setup point. This is possible be-

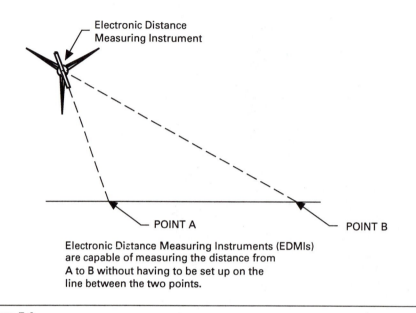

Electronic Distance Measuring Instruments (EDMIs)
are capable of measuring the distance from
A to B without having to be set up on the
line between the two points.

Figure 5-4.

cause the EDMI can determine the distance between two points without the instrument having to be located on the line between the points. See Figure 5-4. It is readily apparent how this feature could be of immediate use in building layout tasks.

5-2   Setting up the EDMI

Using electronic instruments in construction field engineering work requires some minor modification of techniques on the part of the field engineer. Generally, the field engineer will find that his or her job is much easier when EDMIs are used.

Due to the wide variety of EDMIs available, any attempt to go into detail about setup and use of each type of equipment in this text would be futile. However, there are some general items which should be considered regardless of the type or manufacture of the equipment being used. Below is a list of considerations and recommendations pertaining to the use of EDMIs in construction layout work.

1. EDMIs use battery packs for power. This is one of the hardest things for the field engineer to get used to when converting from conventional surveying equipment to EDMIs. Battery packs must be fully charged and

checked before they are taken to the field. It is recommended that a spare battery pack be kept charged and taken to the field as a precaution. Most EDMI battery packs are custom-made for a particular instrument and are not readily available except from the manufacturer, so spares should be kept on hand.

Most EDMIs are capable of running on an automobile battery either through direct connection to the battery or through the vehicle cigarette-lighter receptacle. It is highly recommended that the necessary cables and transformers that would allow the EDMI to be operated off an automobile battery by direct connection be taken to the field. This will allow the field engineer to remove the battery from a vehicle and use the EDMI anywhere on the job site with only minor inconvenience. This is certainly preferable to delaying the field engineering work while battery packs are charged or replaced.

Also, most EDMIs can be supplied with an alternating current (AC) converter. This allows the field engineer to use any convenient 110-volt 60-cycle outlet to power the EDMI. However, convenient power is not always available on the job sites, so relying solely on the AC converter is not recommended.

2. Depending on the complexity of the equipment and the requirements of the job, the operator may have to enter several pieces of data into the instrument prior to or during the survey. Data may have to be read from the display as well. This requires great care as the on-board computer on the instrument can only deal with the information it is fed. Common mistakes are:

Confusing meters with feet. Most instruments allow the operator to select either meters or feet for output and input data registration. Do not set the instrument on meters if feet are to be entered. Conversely, do not read feet from the output display if the machine is actually displaying meters.

Confusing gons with decimal degrees. Both 360 degrees and 400 gons represent a complete circle. An angle expressed in decimal degrees and one expressed in decimal gons may appear to be almost identical. Do not get them confused.

Using incorrect barometric pressure readings. Some instruments require that current local barometric pressure readings be entered to obtain a higher degree of accuracy. Usually, barometric pressure variations will not have a significant effect on accuracy over relatively short distances (200 to 300 feet). However, if there are significant changes in altitude from one point on the traverse to another, or if a major weather front passes through the area, some significant variation in measured distances will occur.

Often, local radio station weather broadcasts are used to obtain barometric pressure readings. However, many radio stations do not obtain their weather reports locally. They are obtained from a central weather-reporting service which may be located many miles from the local station. While this is fine for general weather information, the barometric pressure readings may have changed dramatically from what was last reported. This is particularly true if warm or cold fronts have recently passed through the immediate area.

The best weather information is available from local airports where the Federal Aviation Administration (FAA) maintains weather-reporting services for pilots. Information from these weather stations is available on a limited basis to anyone who calls. The numbers for local FAA weather-reporting services are usually listed in the phone book under the heading of U.S. Government Offices, FAA Flight Service Stations.

Other sources of barometric pressure readings and other weather information are the National Oceanic and Atmospheric Administration (NOAA) weather-reporting stations. These stations continuously broadcast recorded regional weather reports and forecasts. Broadcasts may be received on the following frequencies: 162.55, 162.40, 162.475, 162.425, 162.450, 162.500, and 162.525 MHz.

In addition to the above sources, barometers and altimeters are available from surveying equipment suppliers. These instruments are quite useful in verifying barometric pressure throughout the day without having to periodically call someone or tune radios for the latest readings.

3. Do not confuse horizontal distance with slope distance. Most of the latest EDMIs will automatically read angles of depression or elevation and automatically calculate the horizontal distance from the instrument vertical center line to the reflector center line. Care must be taken to properly set or read the instrument to determine which value—slope or horizontal—is needed.

4. Remember that instrument height, reflector height, and benchmark elevations are handled in much the same way as with conventional transits or levels.

5. Regular calibration of the instrument is recommended. This usually involves sending the instrument to the manufacturer or to a locally authorized repair center for the work to be done. Calibration is absolutely necessary if the instrument is known to have been knocked over or if the instrument has sustained some other shock or damage. A course for checking instruments should be established near the office or job site and should be used regularly. See Section 6-2 for information on setting up such a course.

If an instrument is to be out of service for several days, arrangements may have to be made to rent another instrument. If this is the case, rent the entire system, prisms and all. Do not try to use parts of one system with another.

6. Use only the accessories designed for the EDMI being used. Attempts to use battery packs or other power sources incompatible with the instrument may result in severe damage to the instrument or the accessory. This applies to prism assemblies as well.

5-3   Setting up the Prism Assembly

Each manufacturer of EDMIs supplies prisms which are recommended for use with their equipment. Manufacturers' operating manuals that are supplied with the systems give the recommended procedures for aiming and aligning the prism assemblies.

Some prisms are designed for use with EDMIs mounted on top of theodolites, while others are designed to be used with self-contained instruments. Some prisms require that a mathematical value (called a prism constant) be added or deducted from the distances measured. With some systems, this may be accomplished by entering the prism constant into the on-board computer on the instrument. Distances displayed or recorded will then be automatically corrected. Using the incorrect prism assemblies can result in significant errors in linear measurement.

In precision linear distance measurement, the center line of the prism assembly must be located precisely over the end point of the line being measured. Therefore, the pole or other assembly on which the prism itself is mounted must be directly over the point on the ground which represents the end of the line being measured.

This is accomplished by (1) holding the prism pole vertical using a level which is attached to the pole, (2) mounting the pole "in" a tripod or bipod, or (3) by using a tripod and a tribrach system.

The first method of holding a prism pole vertical using a rod level is similar to the method used with a regular rod. The bubble in a bull's eye level mounted on the prism pole is centered, the instrument operator notified, and the reading taken and recorded. In rough terrain or in windy conditions, holding the pole vertical for a sufficient length of time for the EDMI to read the reflected beam may be difficult. A pair of broom handles, range poles, or other suitable items can be used for braces to overcome this problem.

Figure 2-17 shows the method of using such a brace arrangement on a surveying rod. The same method can be used on a prism pole.

The second method of holding a prism pole vertical is mounting it "in" a tripod or bipod. The tripods and bipods supplied for holding prism poles are designed so that the pole fits securely through the top of the tripod or bipod. The point of the prism pole extends to the ground and rests on the end point of the line being measured. Some field engineers prefer the bipod, because it is somewhat easier to plumb the prism pole. Using the bipod, the prism pole actually becomes the third leg of a tripod. However, it is not quite as stable as a tripod. Figure 5-5 shows prism poles and a prism pole tripod.

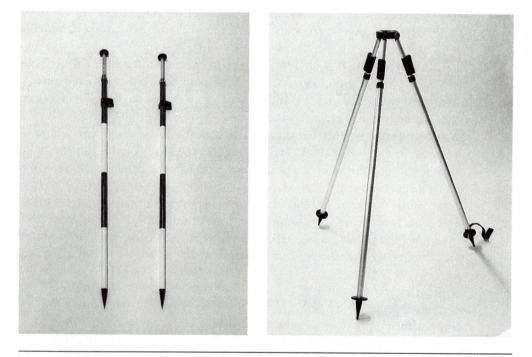

Figure 5-5. Prism poles and prism pole tripod. (Photo courtesy of Sokkia Corporation.)

Figure 5-6. Tribrach. (Photo courtesy of Sokkia Corporation.)

The third method of maintaining the prism in a position directly over a point is to use a regular instrument tripod, a tribrach, and a mounting base for the prism assembly. The tripod is erected, leveled, and centered as closely over the point as possible. The tribrach is then mounted on the tripod.

A tribrach looks like a transit or theodolite but has no uprights, circles, or telescope. In reality, it is the base of a transit or theodolite, with leveling screws, level vials, and optical plummet. See Figure 5-6. The tribrach is leveled and centered over the point in exactly the same manner that a transit or theodolite would be leveled and centered. Review Chapter 3 for details on instrument leveling and centering.

Next, a base fixture is inserted into a recessed area in the top of the tribrach. This recessed area is designed and machined to force the fixture to center itself on the vertical axis of the tribrach. This base fixture is fitted with the prism assembly and, perhaps, a target for visibility. The vertical center line of the prism, target, and base assembly coincide with the center of the tribrach which is centered over the point. Most base fixtures with the prism and target attached can be rotated through 360° on the vertical axis and the target and prism can be tilted several degrees on their horizontal axis.

The prism must be aimed at the instrument as closely as possible to prevent errors. To accomplish this alignment, most prism assemblies have an aiming device built into them or they may be equipped with an accessory aiming device. Using these devices will assure the maximum accuracy in reflecting the beam back to the instrument.

In all cases, the prism should be kept as close to the ground as possible. This reduces the amount of error if the vertical axis is slightly out of plumb. Some manufacturers provide very short prism poles to aid in this effort. Of course, this practice is not always possible if the traverse is through brush or tall grass.

Some field engineers use the technique of locating a point with the prism conventionally mounted on the top of the prism pole. Once the point is located and if the line of sight is clear, the prism pole holder inverts the pole, setting the prism directly on the mark or point at the end of the line being measured. Minor corrections in the location of the point or mark may be made at this time. Of course, not all prism assemblies lend themselves to this technique.

Some EDMIs are adjustable in terms of sensitivity and range. It is possible to turn the sensitivity down on some instruments and "track" a prism as it is being moved from place to place. This is especially helpful in topographic work where tracing an elevation is required. Once the prism pole holder has located the general area where a more-precise reading is to be taken, the instrument operator can change the instrument to a non-tracking mode and to a more sensitive scale and take a reading. The operator may then direct the prism pole holder to adjust the position of the prism pole until the required point is reached.

All manufacturers of EDMIs, prisms, and other accessory items publish owner's manuals and supply them with their equipment. Also, most of the manufacturers will provide classroom or field instruction and seminars on the use of their equipment. It is the responsibility of the field engineer to avail himself or herself of these training opportunities.

Many field engineers produce their own equipment check-out and operation check lists based on the manufacturers' manuals. These check lists are abbreviated but complete, much like those used by pilots for checking out their airplane and preparing for takeoff. When these check lists are used consistently, errors, forgotten equipment, and other embarrassing and costly errors can almost be eliminated. Using the check list enables the field engineer to concentrate on other aspects of the job without having to clutter his or her mind with details about the equipment and its operation.

Some field engineers take the time to lay out a base line with an offset point near their home office or on a long-term job site using a highly accurate tape and a freshly calibrated EDMI. Periodically, perhaps once every two weeks, the field engineer will take a few minutes, set up the EDMI, and shoot the base line and the offset point. A quick observation of the results will tell if the equipment is working properly in all axes. Section 6-2 outlines this procedure and Figure 6-1 shows a typical instrument check-out course.

5-4   Anticipating the Future of EDMIs

Many authorities feel that the field of electronic measuring in surveying and construction is just being explored. There are currently systems available on the market that read the position of the instrument on the earth's surface by reference to a satellite. Some manufacturers are reporting accuracy to within inches using these systems.

Many believe that in the not-too-distance future, satellites will be used to read linear distances, angles, and elevations with levels of accuracy greater than is possible with EDMIs currently available. Perhaps the field engineer of the future will only rarely need to go to the field to do the layout work on job sites because most of the work will be done by satellite communication.

Chapter 5 Review Questions and Exercises

1. How do EDMIs compare to radar and sonar systems?
2. Describe the development of the EDMI as it progressed through the various "generations."
3. List some of the advantages of using EDMIs on construction sites.
4. What are some of the basic cautions that must be observed when using an EDMI?
5. There are single- and multiple-prism assemblies available for use with EDMIs. When is the use of single-prism assemblies recommended? Multiple assemblies?
6. What is a tribrach? How is it used with a prism assembly and for what purpose?

CONSTRUCTION SITE PLANNING, LAYOUT, AND CONTROL

6

BUILDING LOCATION AND STAKE-OUT

6-1   Introduction

Engineers and architects often spend months and sometimes years planning a structure. Every detail is planned, from the physical dimensions of the framework to interior and exterior finishes. In most cases, permits which grant the owner the right to build must be obtained from numerous government agencies. Representatives of these government agencies study the plans in great detail and will only grant permits to build if certain guidelines are followed. Periodically, inspections are performed to ensure that the original plans are being followed.

Along with the careful structural and aesthetic planning, the placement of the structure on the property is also carefully planned. Thought is given to codes, regulations, parking, drainage, and numerous other factors that govern the placement and use of the structure. If the structure is not placed as indicated on the approved plans, the permit to build has been violated and the structure may have to be torn down or moved.

In an attempt to eliminate the possibility that a structure may be placed in the wrong location on a job site, many lending institutions, city building code departments, architects, and engineers require that a professional surveyor verify the position of the structure early in the construction process. However, on many projects no outside surveyor is consulted, so it becomes the responsibility of the contractor and the field engineer to ensure that the structure is not only properly built, but that it is placed on the job site in precisely the location indicated on the approved plans.

6-2   Establishing Dimension Control Program Standards

It is the responsibility of the field engineer either to choose or to assist the project manager in choosing the various dimension control methods which will make up the system to be used on a particular construction project.

Some construction companies have developed and standardized their own systems and will require all of their field engineers to use the company-approved procedures and standards. In other cases, the client may have an engineering staff that has developed certain requirements for dimension control. Often, the client will require the contractor to adopt the client-designed system. This enables the client's representatives to more easily check dimensions during the duration of the project.

However, in the absence of a pre-existing set of company or client standards, the field engineer must assume the responsibility for establishing procedures to be followed to ensure a high-quality field engineering program. In establishing a quality field engineering program, consideration must be given to all items connected with field engineering activities. This extends to equipment purchasing and maintenance, continuing education and training, and establishing measurement and layout standards.

The following paragraphs outline many of the points which must be considered. Specific recommendations and suggestions are the result of many years of accumulated experience on the part of many field engineers and project managers.

Instrument Purchasing, Care, and Maintenance

The general rule is to purchase only high-quality instruments. This rule should apply equally to the complicated and expensive electronic distance measuring equipment and to the simplest of accessories. Quality, however, does not mean "to an excessive degree." Purchase the equipment that will meet the minimum requirements for the greatest percentage of the work to be done.

For example, you might wish to purchase a simpler and less-expensive 10-second-reading electronic instrument for daily use. You could then rent a 1-second-reading "total station" instrument when the occasional job requires it.

Some field engineers wish to purchase only one brand of equipment from one supplier because of price, service, familiarity, and so forth. Others will purchase from at least two sources. There are two principal reasons for purchasing from two or more different manufacturers. First, in case one company goes out of business, the entire fleet of instruments is not left without manufacturer support. Second, many field engineers prefer to use one instrument on the initial layout and then use a different instrument to double check the first process. Many prefer that the second instrument be of a different manufacturer from the first to prevent the possibility of a recurring instrument error.

Immediately after the purchase of surveying equipment, the field engineer should fill out and return all manufacturer's registration cards. A file should be created for each instrument. The serial numbers, and all invoices and certificates pertaining to the instrument, as well as records of calibration work should be kept on file. It may be necessary to report the serial numbers of instruments to the company comptroller or other official in charge of company inventory. This is done for the purpose of internal control of company assets and for insurance purposes.

It is also a good idea to mark the instrument with the name of the company. This must be done carefully to ensure a permanent mark and to avoid damaging the instrument. Consult the manufacturer for advice on the best method and location for identifying inscriptions. Many police departments will loan electric vibratory engravers for the purpose of marking valuables. Use caution when using these vibratory engravers on surveying instruments. The vibratory action may damage the internal parts of the instrument.

A good never-wavering program of equipment maintenance is mandatory. The manufacturer's suggestions should be understood and followed. This will include periodic maintenance of the instrument by the manufacturer or an authorized repair facility.

Periodic maintenance by the manufacturer will accomplish two things. First, it will ensure that the instrument is capable of doing what you purchased it to do when you need it to do it. This will often prevent costly job delays and unwarranted expense. Second, if faced with a legal problem arising from what may be perceived to be inaccurate field measurements, the maintenance record of the equipment may be invaluable in proving that sufficient care was taken on the part of the contractor to prevent such an occurrence.

Transportation and storage of equipment must be considered as well. The manufacturer's manuals will usually contain suggestions that are helpful, but much of the responsibility in this area is left to the common sense of the field engineer. Perhaps the greatest problems are finding a storage area that is (1) secure from theft and vandalism, and (2) protected from the elements.

Storage of surveying equipment and preventing its theft is often a problem on the job site due to the lack of secure and weather-tight storage facilities, particularly in the early stages of site development. Portable buildings or trailers that may suffice for temporary offices and tool storage facilities may not be suitable for storing expensive and delicate surveying instruments. The field engineer may have to transport the instruments back to the home office, to another job site, or to his or her personal quarters to ensure security.

Thought should be given to the advisability of leaving the instruments overnight in a car, van, or truck. Field engineers often must work out-of-town and stay in local hotels or motels for short periods of time. It is suggested that the more-expensive and delicate instruments be brought into the field engineer's room at night or checked into the hotel's secure storage area.

When the instruments are in use on the job site, the field engineer should never let the instrument out of his or her sight. There is often a lot of activity on the job site and many vehicles may be coming and going. It is quite easy for a thief to pick up an expensive instrument and carry it away in a brief moment. If an instrument must be left in place for an extended period of time, the field engineering crew must take turns "guarding" the instrument.

Periodic Checking of Instruments

It is recommended that the field engineer establish a test course on which a biweekly check of the instruments can be conducted. This involves no more than establishing a line on fairly level terrain approximately 200 feet in length, with monuments at either end and one in the center. In addition, one offset point should also be established near one end of the line. See Figure 6-1. The layout work for this test course should be done with the highest-precision instrument available, preferably a freshly calibrated one.

Levels, transits, theodolites, and electronic distance measuring equipment can all be checked on this course. It is suggested that all instruments in use in the area be checked on this course every two weeks or more often if conditions warrant. This test course should also be used to check rental instruments, even those supposedly having been recently shop calibrated.

NOTE

This course is not intended to replace periodically scheduled shop service and calibration. It is intended only as a tool to prevent the continued use of an instrument that may have been knocked out of calibration without the knowledge of the field engineer.

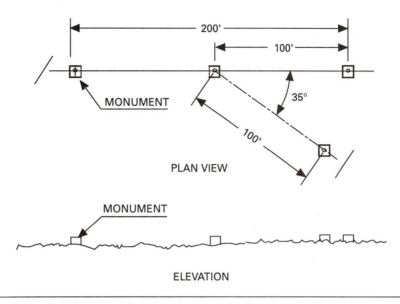

Figure 6-1. Typical instrument check-out course.

Tapes can also be checked on a marked course. They should also be checked against an invar tape. In many instances, tapes will become stretched without the knowledge of the field engineer. Usually, it is more economical to discard the stretched tape than to try to correct it.

Periodic checks of thermometers, scales, and altimeters must be done as well. In some cases, this can be accomplished by the field engineer using a proven instrument for comparison, but in other cases, the instrument must be sent to the manufacturer for calibration.

Selecting Instruments to Use for Specific Purposes

The general rule here should be to select the least-complicated and least-expensive instrument which will do the job. However, some jobs should only be done with specific instruments. For example, laying out a network of control points on a large job site should only be done with an instrument capable of reading to within 1 second of arc and with a corresponding level of accuracy in linear measurement. Inside the network, instruments capable of reading to within 5 or 6 seconds will probably suffice. At individual structures, instruments capable of reading to within 1 minute of arc may be sufficient.

Using a guide similar to the one just described, the field engineer can establish an "order" of accuracy for all job-site dimension control and layout activities.

Tolerances

Just how accurate in building layout and dimension control must we be? Obviously, the client wants the structure to be as perfect as possible, but the client also knows that each restriction or tightening of the standards costs additional money. Therefore, some happy medium between quality and cost containment must be reached.

Several factors will dictate the level of accuracy required on a specific job site. Unfortunately, there has been no one specific standard published in the United States that addresses this problem. The accuracy standards on job sites are a composite of building specifications, organizational recommendations, and manufacturers' specifications. For example, the architect may have written dimension requirements into the specifications that do not correspond with those of the American Institute of Steel Construction (AISC), which may have published similar but different recommendations. Additionally, the manufacturers of certain building components or mechanical systems often require that openings and mounting points for their products be constructed and positioned in accordance with another set of specifications.

These different requirements make the layout work of the field engineer quite difficult. He or she often does not know which instruments to select for use in a given situation or how much time to spend perfecting a particular measurement.

One standard which is gaining some recognition in the United States is the International Organization for Standards publication 4463 (ISO 4463) or British Standard 5964 (BS 5964). This standard establishes limits for primary and secondary control network systems as well as standards for layout points for individual structures within the control network. The ISO 4463 document may be referenced in the specifications of a project, thereby making it the standard for tolerances for that project.

NOTE

Portions of the ISO 4463 document are included in Appendix E.

The ISO 4463 document sets standards by establishing limits between calculated values and actual measured values. This applies to both distances and angles.

EXAMPLE

The standard sets the maximum allowable difference between the calculated distance and the measured distance between primary control points as:

+ or − 0,75 $\sqrt{L}$ mm   where L is the distance in meters

Restated, this simply means that the maximum allowable difference between calculated distances and actual measured distances must be no more than 75% of the square root of the measured distance in meters expressed in millimeters. (In parts of Europe, the comma is used as we use a decimal point in the United States.)

For example, if the calculated distance between primary control points is 36 meters and the measured distance is 36.003 meters, would this be within the limits set by the ISO 4463?

To calculate, plug the values into the formula above.

Allowable difference = .75 × $\sqrt{36}$ meters expressed in millimeters

Note: Convert to mm by dividing by 1000.

$$(.75 \times 6) \div 1000 = 5 \text{ mm}$$

Since the difference between the calculated and the measured distances was .003 m (3 mm), the difference does not exceed the limit and is acceptable under the ISO 4463 standard.

A similar standard is used for angles. It is stated as:

$$+ \text{ or } - \ 0{,}05/\sqrt{L} \text{ gon}$$

Restated, this means that the maximum allowable difference between calculated angles and measured angles must be no more than .05 divided by the length (in meters) of the shortest line associated with the angle. This is expressed in gons. A gon is a European angle measurement equal to .9 degrees. (There are 400 gon in a circle.)

In U.S. terminology and using degrees instead of gons, this formula would be:

$$+ \text{ or } - \ .045/\sqrt{L} \text{ degrees}$$

For example, if the calculated angle between primary control points is 35°52'10" and the measured angle is 35°52'18" and the shortest leg associated with the angle is 48 meters, is the 8" difference within the tolerances set by ISO 4463?

To calculate, plug the values into the formula above.

$$\text{Allowable difference} = .045/\sqrt{48}$$
$$= .006495° \text{ or } 23.38''$$

The observed difference of 8" is well within the allowable difference of 23.38".

Another often-overlooked item which should be addressed when establishing a set of standards is the subject of rounding after calculating. Several methods can be adopted, such as carrying to 7 decimal places and rounding to 5, carrying to 5 and rounding to 3, or perhaps truncating at a certain point. Again, there are no established rules, so it will be the responsibility of the field engineer to choose a method and stick with it. However, most field engineers seem to habitually carry to 5 decimal places and round to 3.

Record Keeping and Reporting

As noted in Chapter 1 of this text, the surveyor's field notebook has long been recognized in the engineering and construction industry as well as the legal community as the accepted method of recording the data gathered during field layout activities. This text presents several examples of field book entries in applicable chapters. Surveying texts will often contain similar examples. Because the field book is so widely accepted as a standard, it should be used faithfully in all field data-gathering activities.

However, the field data book may need to be supplemented with other documentation. This may take the form of time cards, daily reports, photographs, videotapes, or reports called in by phone. The kind of reports required largely depends on the way a particular company collects its cost accounting and payroll data. It may also depend on whether the field engineering crew is paid on a salary basis or by the hour. In some cases, the field engineer's activities are reported by the project manager and no further reports are required of the field engineer.

Considering the variables above, the field engineer must see to it that, one way or another, a record of the activities of the field engineering efforts are reported to the company. This ensures that field engineering activities may be properly logged and charged to the appropriate job by the cost accounting department. This activity-reporting practice will also assist the payroll department in issuing timely and correct paychecks to the field crews.

In addition to field notes and time reporting, the field engineer should familiarize himself or herself with the purchasing and inventory standards of the company. This will aid the field engineer in knowing how to approach the purchase of the supplies and equipment necessary to operate on a daily basis. Files should be kept on all major pieces of equipment purchased. Invoices, manuals, and copies of all paperwork pertaining to instruments should be kept in the equipment file. It is also a good idea to photograph the more-expensive items of equipment and keep the photos in the equipment file.

It should also be mentioned while on the subject of photographs that there are many occasions in which it may be a good idea to photograph some aspect of the job to document a specific condition or event. Many project managers have professional photographers take photos periodically to document progress. Also, many project managers take their own photos of the job. If there is any doubt as to the necessity for photographing some aspect of the project, the field engineer should consult with the project manager to resolve the issue.

While the practice is not mandatory, most project managers keep a daily diary of project activities. Many simply record their diaries on small tape recorders as they drive home each day. Others keep extensive written notes. These daily diaries have served as legal documents in many cases and have often saved many thousands of dollars in charges and legal fees. It is strongly suggested that the field engineer adopt this practice as a part of the standard record-keeping requirements. This is particularly important if the field engineer is moving from job to job on an almost daily basis.

Field Markers

As a part of the field engineering standards, the color coding and the method of making notes on field markers should be specified. The color code covers paint applied to monuments, hubs, or other markers, ribbons on stakes, flags on wire markers, and stake brushes. The notes placed on stakes contain information on elevations, cuts, fills, offsets, directions, and so forth.

There is no rigidly defined standard in existence at this time covering color coding of field markers; however, some organizations have recommended that their members use certain color codes and notes on specific types of work. Appendix F is a consolidation of the color code recommendations of several organizations, both public and private.

The locations for job-site placement of the various markers depends en-

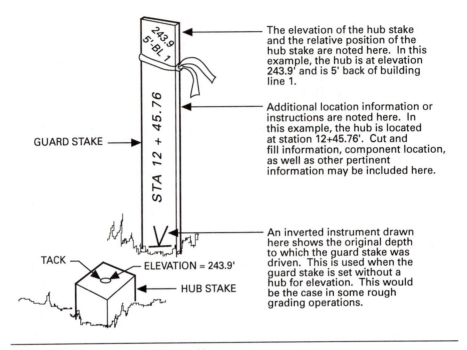

The elevation of the hub stake and the relative position of the hub stake are noted here. In this example, the hub is at elevation 243.9' and is 5' back of building line 1.

Additional location information or instructions are noted here. In this example, the hub is located at station 12+45.76'. Cut and fill information, component location, as well as other pertinent information may be included here.

An inverted instrument drawn here shows the original depth to which the guard stake was driven. This is used when the guard stake is set without a hub for elevation. This would be the case in some rough grading operations.

GUARD STAKE →

243.9
5'-BL 1

STA 12 + 45.76

TACK

ELEVATION = 243.9'

HUB STAKE

Figure 6-2. Typical guard stake markings.

tirely on the nature of the project, terrain, other work in progress, sequence of work, local customs, and many other factors. It is beyond the scope of this text to attempt to define a set of rigid rules in relation to the location of markers. Recommendations are given in the appropriate chapters concerning specific project activities.

Notes placed on stakes are often confusing to anyone other than the person who laid out the job. As with color codes for markers, some organizations are attempting to standardize stake markings. Figure 6-2 shows some suggested markings. Others may be developed and included in the standards for the field engineering operation.

Computer Hardware and Software

The field engineer must be aware of the computer hardware and software available to assist in the field engineering efforts. If presented with the responsibility of acquiring either computers or software packages, the field engineer must do a considerable amount of research before making a purchase decision.

Some points to consider are: (1) Is the computer system compatible with existing computers in use within the company? (2) Is the software compatible with existing hardware? (3) Is the system easy to use or will it require extensive training? (4) Is there supplier support available? (5) Will it do the job which needs to be done?

Obviously, these questions cannot be answered quickly and easily. The field engineer must consult with others in the business, attend trade shows, and read extensively in the trade journals to be able to intelligently buy computer equipment. Belonging to a trade association and attending its meetings is recommended as a source of information on this and other subjects.

Selling the Field Engineering Standards Package

Once the standards for quality field engineering are established, the job remains to sell the concept. First, the proposed standards should be organized into a printed and bound volume, which should then be presented to the supervisor of field engineering activities. The supervisor should be asked to review the standards to see if they conform to company policy.

After review, several copies of the standard should be produced and distributed to appropriate offices within the company. Copies should be distributed to all project managers with the request that the system be explained to all appropriate subcontractors.

The field engineer should be available to attend all pre-job conferences. The field engineer should be on the agenda to present the concepts of the field engineering standards to the subcontractors who will be affected by these standards.

The Team Effort

The field engineering crew is just like any other crew. To get the job done, the crew must work as a team and the lead field engineer must set the example as a team leader. Some say that leaders are born, not made. However, there are certain techniques which should be used when working with people in a team situation which are almost universal and can be learned. Again, the professional organizations within the construction industry can be of assistance in providing leadership and supervision classes.

There are two distinct principles which can be applied to the operation of the field engineering efforts and should be made a part of the standards of operation. (1) The crew leader should never ask another crew member to do something he himself or she herself would not do or has not done, and (2) the supervisor should rotate the individual assignments among all crew members on a regular basis.

The crew members quickly learn that the leader has the experience to do each job, has done it before, and will do it again. This means that there is no indispensable crew member. The crew members usually have more respect for the leader who has actually done all the jobs before becoming a supervisor.

By rotating crew assignments, the individual members know they will not be stuck with undesirable duties indefinitely. This does more to improve attendance than any single thing a supervisor does. This also provides the opportunity to cross-train the individual crew members in all aspects of the crew's responsibility. This ensures continuous operation if a crew member is forced to be absent for some reason.

Of course, there are hundreds of additional supervisory points that could be considered, but the two items above should be a part of the written standards of a quality field engineering program.

Continuing Education and Training

Technology is changing rapidly, as are project requirements. New equipment is being introduced every day and competition is growing stronger.

It takes a concentrated effort to keep up with the latest innovations in the industry. One of the best ways is to join local trade organizations and attend their meetings. Most major construction companies belong (as a company) to their local and national trade organizations. These organizations often provide classes in a multitude of subjects that are of interest to contractor employees.

A second way to keep up with the latest innovations in the business is to subscribe to the trade journals. Often, these journals are available free to those in the business, or copies can be obtained at the contractor's home office or at local libraries. Fill out the reader service cards found in the magazines or call the 1-800 numbers listed to request that the latest literature be sent to you.

Another way to keep current is to visit local supplier show rooms and trade shows. Most of the local suppliers of surveying, engineering, and architectural supplies will know when the next nearby trade show will be held. Complimentary tickets may often be obtained from these dealers. Once at the show, visit with the dealer representatives, get to know them, and get on their mailing lists for the latest brochures describing the latest equipment.

Local colleges and universities often offer courses and seminars on various subjects which are applicable to the field engineering staff. If none are currently offered, talk to the local college officials. They might be able to organize such a class if the need is called to their attention.

Establishing continuing education requirements and requiring attendance at some trade association meetings or shows should be seriously considered as a part of the standards of a quality field engineering program.

6-3   Verifying the Project Location, Boundaries, and Benchmarks

Construction documents for both large and small projects will usually have been developed by many different individuals over a considerable period of time. In these situations, the plans and specifications are sure to contain a few mistakes. Some field engineers take the position that no dimensional information on the prints or in the specifications is to be trusted or used in the construction process until they personally check it out. This is not meant to be derogatory or in any way critical of the architects and engineers who have produced the plans and specifications. This is just the field engineers' way of stating that if a mistake has been made, they intend to catch it before it delays the project or costs someone some money. It is a matter of pride in workmanship and in their profession.

Verifying the Job Site

As ridiculous as it may sound, it is not only possible to build a structure in the wrong position on a job site, but there have been cases in which a structure was built on the wrong job site. This can happen when circumstances such as ambiguous addresses, incorrect street names, and careless pre-job planning come together on one project.

Often, the field engineer is the first employee of the general contractor to arrive on the site of the proposed project. Engineers, architects, surveyors, and others will have visited the job site during the early planning stages, but these visits may have been made months prior to the actual contract being awarded. It is the responsibility of the field engineer to identify the job site, locate and mark the corners, boundaries, and benchmarks, and place the initial stakes that will guide the erection of the structure.

However, before beginning the initial site layout work, the field engineer must first ensure that the proper piece of real estate where the structure is to be placed has been identified. In both developed and undeveloped areas, it is often quite difficult to identify a particular parcel of land. To prevent an embarrassing and costly mistake, the field engineer must identify, without the slightest doubt, the proper location for the proposed project. The following is a list of things the field engineer should do to ensure that the proper piece of real estate has been located.

A. Locate the legal description of the property. Often, the legal description will be found on one page of the site plan or near the front of the specification book or project manual. Without referring to the plot plan, use the legal description to draw a rough map of the property. (Refer to Section 1-6.) Compare your map to the plot plan and to the actual property where the building is to be built. Are they identical or close enough to eliminate any doubts that the proper piece of real estate has been identified?

B. Check the plot plan for landmarks such as street intersections, utility service locations, railroads, fire plugs, and other permanent or semi-permanent items. Compare the area around the job site with the plot plan. Are the landmarks in position and are they in the proper locations relative to the plot plan?

C. Check the names of adjacent streets and nearby highway numbers. Do they match the plans?

D. Locate the control lines or control points from which the plot plan was drawn. Measure along the control lines to identifiable points on the property. Do the distances match the plot plan? Does the scale appear to be proper?

E. Compare the direction indicator on the plot plan with the actual directions at the job site. Use a compass to verify the directions on the job site. Does the compass agree with the direction indicator on the plot plan?

F. Observe the area for several blocks around the job site. Is there a similar area nearby? Consult a street map of the area. Are there any other streets in the area with the same or similar names? (Example: Maple Street, Maple Drive, Maple Road, Maple Circle, North Maple, South Maple, and so forth.) In metropolitan areas, check to be sure that the location is in the proper city. It is possible to cross into an adjacent suburb without knowing it.

G. Try to locate the original surveyor's markers. Often, a recent survey will have been performed when financing for the proposed project was arranged. If the surveyor's markers can be located, compare their location with the legal description, your map, and the plot plan.

If there is any doubt that the job site location is not the one described in the plot plan, the field engineer should notify the project manager or other company officials immediately. The field engineer should document items that have caused this question to be raised, and should be prepared to present the documentation to the proper authorities.

Verifying the Project Boundaries

Once the field engineer is satisfied that the proper site has been identified, the decision must be made about whether or not to identify and mark the property corners and boundaries. This decision will depend on the nature of the project and whether construction is to be controlled by or extended to the boundaries.

On most job sites, the field engineer must measure inward from the property lines to establish the location of the structure, so identifying the exact location of the boundary lines is mandatory. Otherwise, it would be impossible to locate the structure in its proper position on the job site.

If construction is not controlled by or extended to the property boundaries, it may still be to the advantage of the contractor to mark the boundaries to prevent inadvertent trespassing onto adjoining property.

To begin the process of identifying and marking the property boundaries, the legal description, the plot plan, and the site itself must be studied. The following is a list of things the field engineer should do to properly identify and mark the job site corners and boundaries.

A. Using the legal description, an attempt to locate the original surveyor's markers should be made. The legal description should indicate how the corners were marked and the date of the survey will indicate when the markers were placed or last located. To locate markers, the field engineer should look for obvious cleared pathways through brush, trees, and weeds. Trees and bushes may show places where low-hanging limbs were cut off to allow the surveyor to sight and measure along boundary lines. The areas around corners may have been cleared in a circle to facilitate instrument setup. Look for the remains of surveyor's tape. This brightly colored tape will last for some time, often for years in some climates. Old fences and lines of bushes or trees will often indicate old property lines.

It is a common practice to mark corners with iron stakes or pipes. Surveyors or owners will often drive these stakes into the ground until the top of the stake is level with the ground. This is to keep mowers or other equipment from hitting the stake, but the practice often results in the stake being covered up. It may be necessary to use a metal detector to locate covered iron markers.

B. If the corner markers are found, examine them closely. Are the iron stakes driven straight into the ground? If not, there is a possibility that heavy equipment operating nearby has disturbed them. Is there any evidence of excavation, utility construction, or other earth-moving activities in the immediate area? Do the markers show any signs of having been tampered with? Are there any dents, gashes, or signs of bending to indicate that the marker has been hit by some type of machinery or tool? Are all markers of the same size and type of material? This would indicate the likelihood of all the markers being set by the same surveying crew. Do all markers appear to be the same age and is their appearance consistent with the length of time they should have been in the ground? If three corner pins are quite rusty and one appears to be almost new, it is obvious that the newer pin has been recently placed. Does the legal description indicate any offset markers? Surveyors will often offset markers to avoid having to cut through thick brush, so care must be taken not to mistake an offset marker for a true corner marker.

When the original surveyor's markers have been located, the field engineer should take precise linear and angular measurements between the markers and compare the results with the legal description of the property and the site plan. If there are appreciable differences between the field measurements, the legal description, or the site plan, the reason for the discrepancies must be determined before further work can be done. Remember that dishonest landowners, vandals, or others may have moved the original markers.

If the original surveyor's markers cannot be found or if the markers appear to be out of position or in any way false, the field engineer should report these facts to the project manager. The field engineer should never attempt to correct the location of what are believed to be out-of-position markers. In most cases, a professional surveyor will be employed to re-mark the property corners. This relieves the contractor of much legal responsibility and ensures that properly located reference points and lines are available to the field engineer. While there is some expense involved in this practice, it could be more costly if the structure were built out of its intended position.

C. When the corner markers have been identified and their position confirmed, the field engineer should carefully clear the area immediately around the marker and take steps to protect it for the duration of the project. In most cases, three or four guard stakes driven into the ground within a foot or two of the corner marker and flagged with surveyor's tape are sufficient. In rare cases, markers which would be impossible or extremely costly to replace, or those with historical significance, should have protective structures built over them and the area enclosed by a fence.

It is a common practice for field engineers to establish an offset marker for use during construction. This is often necessary when the original marker is in an awkward or inaccessible position and cannot be effectively used for layout purposes. If offset markers are established, care must be taken to place these markers precisely. Also, allowances for the offset distances must be made in all measurements taken during the layout procedure.

6-4   Staking for Clearing and Grading

Construction activities on most projects begin with preparing the job site for excavation. In some cases, demolition of existing structures and the removal of debris is required. In other cases, clearing of brush and trees and the moving of vast quantities of dirt must be completed before the initial foundation excavation can begin. These activities are referred to as "site clearing and grading."

Site clearing and grading may be done by a subcontractor or the general contractor may choose to do it using his own forces. Whether working for the subcontractor in laying out the initial clearing and grading work or working for the general contractor in checking the work in progress, the field engineer is the person usually responsible for measuring and marking the area, indicating the limits of the work to be done.

Preparation for Clearing

Before beginning the layout work for site clearing and grading, the field engineer should consult with the project manager, the chief estimator, or other company officials to see if there are any special provisions connected with the site work. Such special provisions may include having to protect historical or archaeological sites, as well as certain species of trees, plants, and wildlife. In some areas, natural waterways may not be disturbed or altered in any way and storm water run-off must be controlled. Special requirements such as these often do not appear on the plans or in the specifications, but upper management should be aware of them.

If special regulations or requirements exist, the field engineer must take the necessary steps to ensure that site layout work is in conformance with these regulations and requirements. Areas or items which are to be protected must be identified, even if experts in archaeology, biology, botany, or hydrology must be consulted. Often, barriers must be built around trees, vegetation, or other items which must be protected. Dams, holding ponds, diversion canals, and retaining walls may have to be built to control erosion and storm water run-off.

After all special requirements have been identified and arrangements have been made to conform to these requirements, the field engineer is ready to begin the site layout work for initial clearing and grading. This procedure is controlled by the nature of the project, the condition of the raw job site, and what the finished job site must look like.

If the project is to extend to the property lines and if all or part of the structure or its support facilities are to reach to the very edge of the property, the field engineer must mark the boundaries in such a way that heavy equipment operators and other workers can easily identify the edge of the property. The usual method is to set line and guard stakes, with plenty of surveyor's tape on them, at intervals along the boundary lines. On some projects, continuous tape or perimeter fences may be required.

To install markers along property boundaries, the field engineer sets up a transit over the corner, sights along the boundary to an adjacent corner, and directs an assistant in setting stakes along the line of sight. The same procedure is followed if an offset boundary line is used. The stakes may be set on the offset line or may be set back on the boundary line depending on the requirements of the job.

The intervals at which boundary stakes are placed depends on the terrain and the visibility in the area. Heavily wooded areas may have to be partially cleared in order for the lines to be marked. In this case, the field engineer may have to be on the job site continuously to provide line for the clearing crew. Remember, lawsuits may result if work is performed beyond the property lines.

If the original staking of the boundary lines can be relied on to remain undisturbed throughout the project, the field engineer may elect to establish construction control points on the boundary line as it is being staked. This procedure is further discussed in Section 6-5.

When the boundaries and areas of special interest have been identified, marked, and protected where necessary, the clearing operations may proceed. Once the site is cleared of trees, bushes, debris, old structures, and so forth, the staking for rough grading can begin.

Preparation for Rough Grading

In preparing to mark the job site for the rough grading operations, the field engineer must again study the plot plan and the specifications. Particular attention must be paid to the existing and finished contour lines for site grading. The student who is not familiar with the drafting conventions depicting contour lines is encouraged to consult Chapter 9 of this text before proceeding.

A general feel for the existing terrain compared with the finished terrain must be developed. The location and elevation of various benchmarks must be identified. In some cases, topsoil must be stripped and stockpiled for re-use in landscaping upon completing the project. Fill dirt may have to be hauled in to fill low areas or excess dirt may have to be hauled away. The ideal situation is to simply rearrange the soil on the job site without having to haul away or haul in soil. This is called a "balanced cut and fill" and it is rarely the outcome on a project.

The grid method (linear measurement along grid lines) of laying out the job site for rough grading is used in the following example. At the present time, the grid method is the most widely used for rough grade layout work. However, electronic equipment and computer programs that lend themselves to radial staking are available and may be adapted to this process. Radial layout is a method of establishing points on a job site by relative angle and distance rather than by linear measurement only. Radial layout is discussed in detail later in this chapter.

Once the field engineer has studied the job site and the site plan and is familiar with the proposed changes in the terrain, the staking for rough grading may begin.

Setting Rough Grading Stakes

The following is a step-by-step procedure for staking the site for rough grading. The fictitious site plan shown in Figure 6-3 is used as an example. Please note that this site plan is abbreviated for clarity. An actual site plan would contain hundreds of detail items which cannot possibly be shown in the relatively small pages of a textbook.

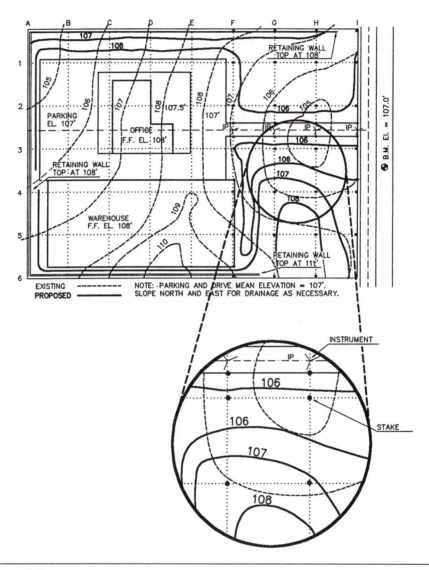

Figure 6-3.

Step 1. In the office, overlay the site plan with a grid. The size of the grid squares depends on the terrain and the level of accuracy needed in the initial rough grade staking. Usually, the rough grading can be done with a grid overlay using 100 ft × 100 ft squares. The dotted lines in Figure 6-3 represent the 100-foot-square grid pattern overlay. Note that the grid lines have been labeled 1 through 6 on the left and A through I across the top. Also, note that the grid squares at the top of the plan are not full size.

Step 2. In the field, choose an initial layout line by observing the terrain and the proposed areas of construction. The idea is to establish this line in a position from which as much of the job site as possible is visible from any position on the line. In the case of the job site depicted in Figure 6-3, an initial layout line running from right to left down the center line of the road would serve quite well if the job site is dry. This line might have to be established farther to the north or south in order to avoid the lower areas in the rainy season.

Step 3. Assuming that the center line of the road in Figure 6-3 has been chosen as the location for the initial grading layout line, the field engineer should set up a transit at the intersection of the initial layout line and the first north-south grid line. This is line "I" and is also the east property line. A backsight on the benchmark across the highway is taken and the "HI" (height of instrument) is recorded.

NOTE

The field engineer may wish to establish a temporary job-site benchmark at this time. This will alleviate the necessity of having to cross the highway to take a backsight reading each time the instrument is moved. A suggested ideal location would be at the approximate geographic center of the job site. See Section 2-9 for information on transferring and establishing benchmarks.

Step 4. The instrument operator should rotate the transit until the northeast property corner is sighted. The assistant should be directed to measure along the property line until the north edge of the proposed road is reached. (A tape or an odometer may be used here for linear measurement. Pacing may be accurate enough in some instances. A quick stadia observation may be used as a check.) Line is maintained by directions from the instrument operator. A stake is set at this point and properly marked and flagged. This point is recorded as a frontsight and the difference between the actual and the proposed elevation at this point is calculated and marked on the stake. This difference is indicated as a cut or fill. See Section 6-2 for stake marking recommendations.

 NOTE

Setting a hub stake is optional. Some field engineers, in this early stage of job-site layout, simply mark the guard stake with the pertinent information and do not set a hub until later, just prior to the fine grading process. Local custom often dictates this practice. Consult the project manager or other supervisors regarding this matter.

Step 5. The assistant should continue measuring along the "I" line to the north, staking each point at which the grid line intersects the property line. At each point the stake is marked with the difference between the actual and proposed elevation and whether this difference indicates a cut or fill.

NOTE

It is a good practice to set initial grade stakes not only at the grid intersections, but also at the point structure, parking lot, or road lines cross the grid lines. A rough placement of structure corners may be required as well. Some contractors will ask that additional initial grading stakes be set at various locations around the job site. The field engineer should consult with the project manager to determine if this is the case. It is much easier to set additional stakes at this time than to return later.

Step 6. The transit is rotated or plunged and the southeast corner marker is sighted. The assistant measures along the "I" line and stakes the grid intercept points as before. The road edge is also staked.

Step 7. The transit is moved to the intersection of the initial layout line and line "H" and the procedure described in steps 3 through 6 is repeated for points on line "H" until line "H" has been "run" from north to south.

Step 8. Continue along the initial layout line until all grid lines have been "run." Figure 6-3 shows the instrument locations and stake locations for the first four lines on this job site.

Step 9. Establish a north-south secondary layout line and repeat the process as described above using the numbered grid lines if necessary. This is usually only necessary in cases of reduced visibility or in cases where the grid size is to be reduced in some areas.

Step 10. Review your work. Get to a high vantage point if possible. Does the layout look right? Sight along the grid lines. Do the stakes line up? Pick a point or two at random and double check the stake information. Does it make sense? Are the proper entries made in the field book? Check the math in the field book.

Step 11. Complete the necessary paperwork, file the field notes, secure the instruments in accordance with accepted procedure, and report the completion of the job to your supervisor.

NOTE

Local customs often dictate how the proposed location of the site for the building, parking lots, and roads are marked. The outline of the building foundation is often referred to as the "footprint" of the building. Generally, for rough grading purposes, the footprint area is marked with stake ribbons that differ in color from the rough grade stake ribbons. The corners of the building footprint are marked, as are additional points on the perimeter and a number of points within the interior. The interior stakes may be located by grid lines or they may be located with reference to the geometric shape of the footprint of the structure. Roads and parking lots are similarly marked.

The field engineer should check with the project manager to determine whether the building pad, parking lots, and road stakes should reflect the finished surface elevation or the elevation of the surface on which select fill material, concrete, or asphalt will be placed. This is dependent on the design of the foundations, the requirements for base preparations for parking lots and roads, and the planned construction sequence.

During the rough grading operations, the heavy equipment operators will bring the building, parking, and road areas that are high down to the finish floor or finished surface elevation or just slightly lower. Any areas which are below the proposed elevation should not be filled unless filled with select fill material and compacted to rigid specifications. This practice prevents the creation of soft areas under concrete slabs poured directly on the ground.

The building, road, and parking areas will subsequently be excavated as necessary, refilled to a specified elevation with compacted select fill material, and the finish surface material applied.

Preparing for Finish Grading

After the select fill material is brought in, spread to a uniform depth, and compacted, the finish grading operation may begin. The finish grading operation is initiated much like the rough grading operation. A base line is established, a grid laid out, and elevations marked along grid lines. The main difference is that the finish grading operation is more tightly controlled than the rough grade procedure.

To begin with, the grid used to lay out the site is smaller. It is usually no more than 50 feet square. In critical areas, a 25-foot or smaller grid may be used.

The benchmarks are read more accurately, usually down to one tenth of a foot. The linear distances are measured with a tape to within a tenth of a foot.

Hubs are set with the top of the hub even with the finished elevation of the base or fill material at that point. This often requires that the area where the hub is to be set be excavated slightly. Traditionally, the hub and the guard stake for finish grading are marked with blue ribbons, blue paint, and blue brushes. The term "blue topping" means to bring the fill material to the finished elevation ready for concrete or asphalt to be applied. A good "blue top operator" refers to an experienced equipment operator capable of controlling heavy earth-moving equipment to such a degree that the finish elevation is reached quickly and efficiently.

Experienced equipment operators can easily control the largest motor graders to within a tenth of a foot of cut or fill. In some cases, an assistant to the equipment operator will run ahead of the machine, read the elevations on the grade stakes, and signal the operator the amount to cut or fill. If grade stakes are clearly marked, closely spaced, and oriented toward the operator, the aid of an assistant may not be necessary.

Some earth-moving equipment is equipped so that the cutting blades are controlled by laser beams. The operator only guides the machine; the depth of cut or fill is controlled by the laser.

On most sites, finish grading is only done on building pads, parking lots, and roads. The surrounding areas of the site are usually landscaped. The elevation of the finished terrain in these landscaped areas is not critical. Often, the rough grading operations are sufficient if the area was not unduly disturbed during construction operations.

One important thing to remember is that the finish grading operation is usually weather dependent. It must be accomplished quickly so that finish materials can be applied before rains can damage the finished areas.

6-5   The Project Dimension Control System

Once the job site is clear and the rough grading operations have been completed in the areas where the structures will be located, the field engineering crew can start the more-precise layout work which will guide the erection of the structures. This layout work is begun by establishing control points from which control lines are extended. The structures are located with reference to these lines and points.

There are two methods used to lay out construction control points. One is the linear method, the other is the radial method. Each method has its advantages and disadvantages. In some cases, a combination of the two methods is used.

Linear Control System

Plans which lend themselves to linear control methods are drawn with the structure lines laid out from the property boundaries and corners. That is, the building lines are placed a certain distance from the property corner measured along and inward from the property lines. Building shapes are generally rectangular and usually run perpendicular to or parallel with property lines.

The linear method of establishing control lines and points is the most widely used at the present time. It works quite well on small job sites and lends itself to the use of conventional surveying and linear measuring equipment for layout work. This method is popular with architects because it does not require extensive on-site pre-planning or layout work. Local governments like the linear method because it readily gives their permit officials information on set-backs, coverage, and so forth.

Using the linear layout method is quite simple in theory. The job-site boundaries and corners are located. Measurements are taken from the corners, offset points established, and building corners staked.

Radial Control System

Plans which can be used with the linear control system can also be used with the radial control system. However, some plans are drawn based entirely on the radial control system. Plans which are designed to use the radial layout method use control points set at strategic locations around the job site to locate the building. The building lines are run between points located by angle and distance from a set of control points. There is little, if any, reference to the property boundaries or property corners.

The radial method of establishing building location on the job site is usually used on larger sites, irregularly shaped sites, and sites where the structures are some distance from the property lines and corners. In using the radial method, one or more control points is established on or near the job site. Measurements for the entire project are taken from these control points. This involves precision angle and distance measurement which, of course, requires high-precision equipment.

Often, the field engineer must use the primary control points to establish a network of secondary control points. The secondary control points can be placed in more strategic and convenient locations for daily use in job-site layout work.

Computer programs are frequently used to aid in the layout procedure when the radial control method is used. The location of primary and secondary control points are entered into the computer, usually by referring to a grid coordinate system. Structure locations are entered as well. The computer will give the azimuth directions and distances from selected control points to the structure corners, center lines, and so forth. All the field engineer has to do is set up an electronic instrument over the specified control point, sight another specified control point, turn the necessary angle to the right or left, measure the specified distance along the line of sight, and drive a stake for the structure corner, offset, building line, center line, and so forth.

Control Points

There are four major categories of job-site control points. These are: (1) Main reference points (sometimes called reference azimuth points or RAPs), (2) primary control points, (3) secondary control points, and (4) building layout points. Some projects will require the use of all four types of control points as well as additional subcategory points. Other smaller and not-so-complicated jobs may require the use of only one or two categories of control points.

Main reference points. The main reference points for a construction project may be located some distance from the project site or they may be located within the site boundaries. This point may be a single point or a series of points. In radial layout procedures, the main reference point can be thought of as the hinge point from which the entire project is laid out. Figure 6-4(a) shows

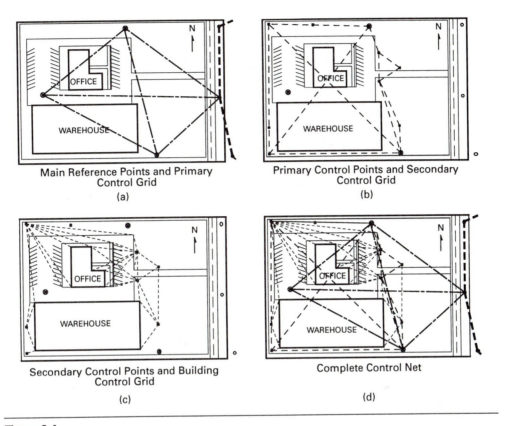

Main Reference Points and Primary
Control Grid

(a)

Primary Control Points and Secondary
Control Grid

(b)

Secondary Control Points and Building
Control Grid

(c)

Complete Control Net

(d)

Figure 6-4.

the main reference points and the primary control points which are established on the job site using the radial technique. Note that only one of the main reference points is used as a "hinge" in establishing the primary control points. If the linear staking method is used, more than one main reference point will usually be used for alignment.

Main reference points are usually located and marked by architectural or engineering firms during the pre-job planning and layout of the project. It is the responsibility of the field engineer to find these main reference points, verify that they appear to be genuine and undisturbed, and use these points to establish the primary control points on the job site from which the layout work will progress.

Primary control points. Once the main control points have been located, the field engineer must establish the primary control point system on the job site. Since these points will be used to control the location of all elements on the site, the most accurate instruments and measurement devices available must be used in the layout procedure. This means using the best electronic equipment to obtain accuracies in the 1:20,000 or 1:50,000 range.

The primary control points are usually marked by building a concrete monument on which the exact point is inscribed. These monuments must be given a high priority level in protecting them from damage during construction. The requirements for deciding on the location and the actual construction details of the control point marking system are discussed later in this section.

Secondary control points. The secondary control points are additional control points placed at various locations within the job site for convenience and to expedite the construction of individual structures on the site. Figure 6-4(b) shows the secondary control grid. Many of the secondary control points within this grid may be destroyed as construction progresses and the particular point is no longer needed.

The accuracy level needed to establish the secondary control points is usually not as critical as that needed for primary control points. Accuracy ranges in the 1:10,000 range is usually more than adequate. The secondary control points may be marked by constructing a concrete monument or by setting hub stakes surrounded by protective laths, posts, or fencing if required.

Building layout points. The building layout points are usually found with reference to the secondary control point system. Figure 6-4(c) shows how the secondary control points may be used to sight the building control points. Note that all major structure corners are visible from one or more building control points. These are the points from which the actual measurements for construction activities will be taken. The points usually identified as building layout points are the corners of the structure and the building lines. These points are usually set back from the edge of the actual structure line so as not to be disturbed by construction activities.

An accuracy level of 1:3000 or 1:5000 is usually considered sufficient in setting the building layout points. The building layout points are usually marked with hubs and guard stakes only, but may be marked with concrete monuments on long-term job sites if subsequent construction activities permit.

The nature and shape of the building will dictate the placement of building layout points and the general conditions of the job site will govern the placement of the primary and secondary control points. Of course, in some circumstances, the primary or secondary control points may also serve as building layout points as well.

NOTE

Many field engineers establish off-site control points which may be used as backups to the on-site points. These points may be such places as a mortar joint on a nearby building, a scratch on a fence across the street, and so forth. It is not uncommon for the field engineer to establish these off-site points and not reveal their location or existence to anyone. This practice is just another backup to locating points on the job site which may be destroyed or disturbed during the progress of the project.

Setting Monuments for Control Lines and Points

In using either the radial or linear control method, establishing proper and accurate monuments is of utmost importance. Locating the control point monuments on the job site should be done in consultation with the project manager. Monuments must be placed in areas where they will be accessible but undisturbed during the duration of the project. The project manager will have a plan of construction in mind and will have the best idea as to where the monuments can be located so they will not be covered up or obstructed by later construction activities. Monuments must not be placed in swampy areas or other areas where the soil is unstable. Ideally, at least two other monuments will be visible from any given monument at all times. This allows a monument to be easily re-established if it is damaged or destroyed.

The geometry of monument placement must also be considered. Chapter 2 introduced the concept of using equal distances between frontsights and backsights as producing the best results in leveling. The same thing applies in establishing control point locations. The field engineer should strive to locate the control points so that triangles drawn between associated points form equilateral triangles. This practice will almost entirely cancel out the effect produced by an instrument which may be slightly out of level calibration.

After the entire grid and all its points have been fed into the computer, the nets must be corrected. Since no survey or layout problem can be done with 100 percent accuracy, deviations discovered must be adjusted. This is usually done by dispersing the error over the entire net by percentages.

For example, if a traverse covered a total of 1080 feet on the plans and the actual final closing measurement was 3 inches over for a total of 1080'3 ", the 3-inch error could be proportionally distributed over each line measured in the traverse. If one line accounted for one third of the total distance, this line would be increased in length by 1 inch. Of course, this may mean that the course must be re-run and the hubs re-set to reflect the corrections.

Once the monument general locations have been selected, the field engineering crew should run a preliminary layout exercise to establish the monument location point more exactly. This exercise should locate the monument center to within 3 or 4 inches of its final precise location. Next, the field engineering crew should excavate a hole 12" to 18" square and about 18" deep at the location just selected. A plywood or 2 × 4 form slightly larger than the hole should be set around the hole and the hole and form filled with concrete. See Figure 6-5. If the monument is to be used for benchmark elevations, concrete re-bar, brass disks, or other metal markers may be set in the concrete to mark the height. The concrete should be trowled until quite smooth. See Section 2-9 for additional information on transferring and marking the elevation on the monument.

Once the monument concrete has set, the transfer of location data is begun. An electronic instrument or a one-second reading theodolite is required to accurately establish precise marks on the control points. The control points, whether concrete monuments or hubs should be color coded with spray paint. The exact line or point on the monument should be marked with an indelible marker and a scratch awl or it should be chiseled into the concrete. Hubs should have a tack set in them at the exact location of the point which is being marked.

Dual Markers

As previously noted, some markers may serve two or more functions. For example, a primary control point marker may be set on an extended building line. This would permit this marker to be used as both a primary control point and a building layout point when necessary.

Often, the contractor may need to establish lines which are offset from but parallel to the building lines. This aids in the installation of steel columns and the building of concrete column forms. (Establishing offset lines for construction purposes is covered in Chapter 8.) It may be possible and desirable to establish the offset points at the same time the building line and corner layout points are established.

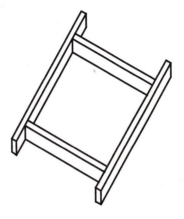

FORM FOR A 2' SQUARE MONUMENT

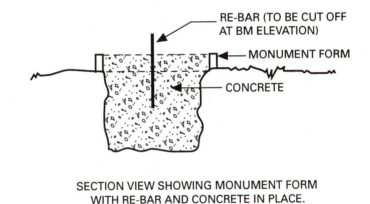

RE-BAR (TO BE CUT OFF
AT BM ELEVATION)

MONUMENT FORM

CONCRETE

SECTION VIEW SHOWING MONUMENT FORM
WITH RE-BAR AND CONCRETE IN PLACE.

Figure 6-5.

Often, the building line and offset marker may be placed on the same monument. This is done by constructing a long narrow monument with the building line reference mark on one end and an offset reference line on the other end. Proper color coding of the ends of the monument is recommended. Figure 6-6 shows the placement of dual-purpose monuments.

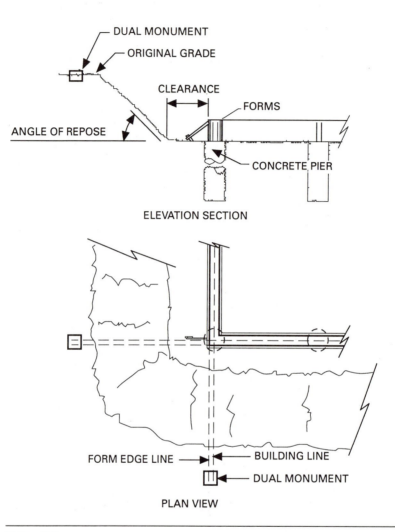

Figure 6-6.

6-6 Establishing Building Corners and Setting Batter Boards

Building corner stakes are rarely set at the building corners. This is because the stakes would be in the way of construction from the very beginning of the project. Building corner stakes are set back some distance from the point where the actual corner of the building will ultimately be. Just how far back the stakes are set depends on several variables. Among these variables are excavation requirements for basements or footings, angle of repose of the soil, subsequent construction activities, and local custom.

If a deep excavation for basements or footings is required, the building corner stakes must be set back far enough so as not to interfere with the excavation activities. The depth of the excavation, the angle of the sloping sides of the excavation, and the form clearance at the bottom of the excavation must be considered. Figure 6-6 illustrates these requirements in profile view. However, building corner stakes or building line stakes may have to be set near the edge of the excavation in order for the field engineer to provide line and grade for foundation construction or pier drilling or pile driving operations.

If a deep foundation is not required for the structure, there may be cases in which a greater-than-normal setback of corner stakes is still required. This may be due to the need to get construction equipment close to the building, set scaffolding, and so forth.

Local custom may dictate the setback distance for the corner stakes. In some localities, a five-foot setback is normal, while in other areas, ten feet is customary. In any case, once the building corner stakes are set, they are marked to indicate the setback distance and direction.

Batter Boards

When the nature of the structure allows, batter boards are used to aid in the excavation and in the setting of the foundation forms. This is usually the practice on smaller buildings. If batter boards are used, some excavation work and foundation concrete work can be done with reference to string lines stretched between batter boards. This relieves the field engineer or others from having to use an instrument to provide line and grade for these activities. The workers excavating and building forms can simply measure off the string lines when they need to check the progress of their work. Figure 6-7 illustrates the batter boards arrangement for a small building. Note that the string lines run parallel to the edge of the foundation.

The elevation of the string lines is optional. Most form carpenters prefer that the strings be set at the same level as the top of the form, others like the strings to be placed a foot or so above the forms. The field engineer should check with the form carpenter lead person or the project engineer to determine their preference.

If batter boards are carefully set and not disturbed, it may not be necessary for the field engineer to return to the job site until the forming is complete.

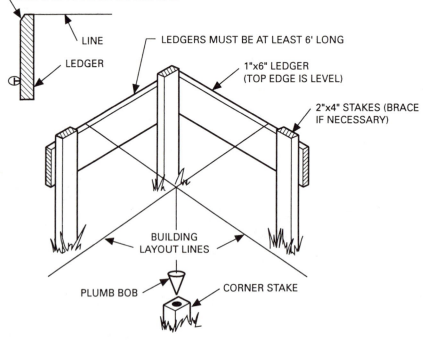

SAW CUT IS MADE IN LEDGER
LIKE THIS TO PLACE LAYOUT LINE

LINE

LEDGER

LEDGERS MUST BE AT LEAST 6' LONG

1"x6" LEDGER
(TOP EDGE IS LEVEL)

2"x4" STAKES (BRACE
IF NECESSARY)

BUILDING
LAYOUT LINES

PLUMB BOB

CORNER STAKE

Figure 6-7. Typical batter boards.

However, a final check for alignment and elevation should be done by the field engineer prior to pouring concrete. In some cases, the field engineer must be present during the pour to verify form elevations and alignment.

Chapter 6 Review Questions and Exercises

1. What are the two main problems encountered in finding a storage area for surveying instruments on a job site?

2. Sketch or lay out an instrument check course. Describe how this course is to be used by the field engineer.

3. What is the ISO 4463? How can it be used in establishing standards for a field engineering program?

4. Why is a color code recommended for field markers?

5. List several personnel management principles that should be practiced by all field engineering crew leaders.

6. List at least three things a field engineer should check to verify that the proper job-site location has been identified.

7. What is an "offset" marker?

8. Should the field engineer replace or relocate missing or disturbed corner markers if he or she is reasonably sure of their proper location?

9. What are some of the precautions that should be taken prior to beginning site layout for clearing and grading?

10. Demonstrate the proper method of laying out a grid for setting grade stakes. Demonstrate the proper method of marking grade stakes for cuts and fills.

11. What is the "footprint" of a building?

12. What is meant by the term "blue topping"?

13. What are the major differences between the linear and the radial methods of site layout?

14. What is an off-site control point? Why is it important?

7

DIMENSION CONTROL DURING FOUNDATION CONSTRUCTION

7-1 Introduction

After the site grading and initial layout work has been completed, preparations for foundation construction may proceed. There are two basic classifications of foundation types used in commercial construction: shallow and deep.

Shallow foundations are used for buildings of one or two stories which are not to be used for heavy manufacturing and which will not have extremely heavy floor loading. In some cases, the ground floor concrete slab is simply poured directly on the ground. The ground will have been compacted and brought to a specified elevation to form a building pad during the fine grading process. Select fill material may have been brought in to complete the final few feet of the pad. Ditches and depressions are dug in the building pad and allowed to fill with concrete during the floor-pouring operation. These thickened areas give the finished floor additional strength.

The columns for the building are set on the surface of the slab (usually over a thickened area) and the entire weight of the building frame is thus allowed to bear on the slab. The use of this type of foundation is recommended for use only in areas where soil conditions and climate permit. This method of construction is called slab-on-grade construction.

The responsibility of the field engineer in laying out the foundation for slab-on-grade construction consists of verifying the elevation of the base material, locating the building corners, setting the batter boards and/or offset stakes, setting elevation indicators for the top of the finished slab, and monitoring the alignment and elevation of the forms.

A deep foundation is a variation of the slab-on-grade and is often used when soil conditions and building design are not conducive to conventional slab-on-grade foundations. In this method, the columns are set on a system of driven piles or cast-in-place piers. The piles and piers extend into the ground some distance. This distributes the weight of the building frame to the piles or piers where it is distributed to the surrounding ground. The ground floor of the building may "float" on the ground independent of the frame or it may be supported by various means.

Most large commercial buildings, particularly multi-story structures, use a deep type foundation but in a much larger scale. Large commercial building foundations are usually begun some distance below the level of the surrounding ground. This allows for mechanical rooms and parking areas to be placed in basements. This practice conserves expensive real estate space, provides convenient services, and is energy saving to some degree.

The responsibility of the field engineer in deep foundation construction consists of laying out the excavation for the lower components, locating building corners and offset stakes, monitoring the placement, alignment and elevation of piers, piles, caps, slabs, and other components of the foundation structure.

NOTE

The specifications for some buildings, local codes, or requirements established by lending institutions may require that the location of the building foundation forms be verified by a registered engineer or a registered surveyor before the first concrete is placed. In these cases, the project manager will usually arrange for this service. It should be noted that this practice in no way questions the skills of the field engineering crew. This is simply a quality-control measure similar to any other inspection requirement.

7-2   Slab-on-grade Foundations

Dimension Control for Slab-on-grade Foundations

Slab-on-grade foundations are laid out using either the linear or the radial staking method, as described in Chapter 6 of this text. Batter boards may or may not be used. The following is a step-by-step procedure for maintaining positive dimension control during construction of a slab-on-grade foundation. This procedure assumes that the building corners have been marked with offset stakes or batter boards in accordance with the procedures described in Chapter 6 and that concrete forms have been erected.

Step 1. Set up a transit over any undisturbed corner control point. This may be an offset corner point or an outside building line point. Sight on the opposite corner control point, then rotate the scope downward and sight along the forms. Note whether the form edge is in the proper relationship to the line of sight of the instrument. Most forms can be held to within plus or minus ⅛" of the sight line. Inform the project manager if the forms are out of line. Be prepared to direct the alignment of the forms.

Step 2. Locate the benchmark used to establish the form height. Verify that the benchmark is undisturbed and have the assistant set a rod on it. Take a backsight reading on the rod set on the benchmark. Have the assistant move the rod to the forms. Locate the top of concrete indicators inside the forms. These indicators will be items such as nails, chalk lines, or chamfer strips. The forms will be filled with concrete to the level of these indicators.

NOTE

In some cases, the forms will be set so that the top of the form is at the same elevation as the top of the finished concrete. In this case, there will be no indicators inside the forms.

Have the assistant hold the bottom of the rod even with the indicator or other form area which will correspond to the top of the finished concrete. Take a frontsight reading on the rod and calculate the elevation of the top of the concrete indicator.

Step 3. Consult the structural prints for the project. Determine the specified elevation of the top of the concrete in the forms just observed. Is it correct? If not, investigate by double checking the prints, the math involved, and the elevation of the benchmarks.

Step 4. Move to an adjacent building corner control point and repeat steps 1 through 3. When the second form line has been verified as correct in relationship to the first form line, straight, and the top of concrete elevation indicators verified, continue around the building until all the form lines have been verified as to position and elevation.

 NOTE

On all square and rectangular areas, always measure from corner to corner both ways if possible. This will give an indication as to whether the layout is out of square or not.

Step 5. Review your work. Make the proper entries in the field notes, secure the equipment, and report to the project manager the results of your work.

7-3   Deep Foundations

The construction of deep foundations requires many more steps than is required for shallow or slab-on-grade foundations. The responsibility of the field engineer is to see that each component of the foundation is constructed on line and brought to the specified elevation. This requires that dimensions and elevations be controlled with varying degrees of accuracy from the time excavation is begun until all foundation components are in place.

There are two methods used to specify the elevations of building components and other pertinent job-site elevations. One method is to use the actual elevations based on mean sea level (MSL). The other method is to arbitrarily select an elevation which will apply to this one job and is not associated with any actual elevations.

Project elevations based on MSL are simply used in a differential leveling operation just as described in Section 2-7 of this text. Backsights are taken on benchmarks of a known elevation and frontsights are taken at other desired locations. The mathematics yield the elevation at the second locations.

Arbitrary elevations are just as easily used as MSL elevations if the arbitrary elevation 0.0 ft point is well below the deepest component of the structure. Problems sometimes arise when the architects specify the arbitrary elevation to be 0.0 ft at the level of the finished first floor. This forces the use of negative numbers for elevations on all components below the finished first floor. Sometimes, the negative signs are missed on the prints and elevations are added instead of subtracted. Field engineers must use extreme caution when working on projects where this system of noting elevations is used.

The preferred method of noting elevations using arbitrary elevations is to set the finished first-floor elevation at 100.0 ft. This will give the foundation components positive elevations and makes calculations somewhat easier.

7-4   Dimension Control of Excavations

In order to lay out the job site for excavation operations and to establish the proper height of forms and other components of the foundation, the field engineer must study the structural prints and soil reports for the project in great detail. Information about the elevation of all foundation components in relation to job-site benchmarks must be determined from the prints. Information concerning the soil that is expected to be encountered during excavation is contained in the soil reports.

Soil Reports

Along with a study of the structural prints, the field engineer must have some knowledge of the soil conditions that will be encountered when excavation begins. This information is contained in the boring logs and soil engineering reports. The boring logs are sometimes included in the prints or they may be found in the specifications. In some cases, the soil reports are in a separate document prepared by the project engineer or architect.

The field engineer will usually be more interested in the boring logs than other parts of the soil report. Boring logs are profiles of soil exploration holes that were drilled at various locations on the job site as a part of early construction planning. The boring logs contain information as to the types and depths of soils, rock, and other materials that were encountered during the boring operation. If a water-bearing strata was encountered, this information is included as well.

The field engineer must know the types of soil that will be encountered in order to properly plan the setback distance for layout stakes, plan for ramps, and allow for stockpile areas.

Different types of soil have different angles of repose. This is the angle between the ground and the sloping side of a mound of soil as it would pile up if poured from a container and allowed to mound up naturally. Of course, moisture and compaction would affect the angle of repose.

When the excavation begins, the sides of the excavation must be sloped. The angle of this slope is dictated by the natural angle of repose of the soil, moisture, compaction, and safety regulations. Excavations in sandy soil have relatively low angle-of-repose slopes while excavations in rocky soil have high angle-of-repose slopes. The field engineer must allow for the angle of repose when placing control stakes.

If the excavation poses any danger to workers from caving, the sides of the excavation must be shored up or they must be sloped back to an angle of repose that is considered safe. Government regulations cover these requirements in detail. The field engineer should consult the company safety representative to clarify the slope and shoring requirements for excavations as necessary.

The field engineer will occasionally discover that if the soil is allowed to seek its natural angle of repose during excavation, the edge of the hole will exceed the limits of the property boundaries. In these cases, shoring or retaining walls must be constructed to allow excavation to continue to its specified depth.

On some projects, the field engineer must lay out areas for on-site storage of excavated material. The minimum area required will be directly related to the volume and the angle of repose of the soil.

Additional Excavation Problems

Occasionally, the field engineer will be asked to establish a system for monitoring the excavation to ensure that it is stable and that the sides are not beginning to "creep" (slide down-hill) and that the surrounding surfaces are not beginning to "subside" (sink). To monitor this situation, the field engineer must set a series of reference stakes along the sides of the excavation and a series of elevation hubs at the top of the excavation. Figure 7-1 shows how these stakes and hubs should be placed.

If any part of the slope moves downward, the stakes will be out of line. This is easily spotted even without an instrument. The project manager will usually have the stakes monitored on a daily basis and can take corrective action to prevent a cave-in.

Monitoring the surface subsidence hubs will, of course, require an instrument. The elevation of the hubs relative to a known elevation on the site is compared to the elevation recorded the day the hubs were set. If some or all of

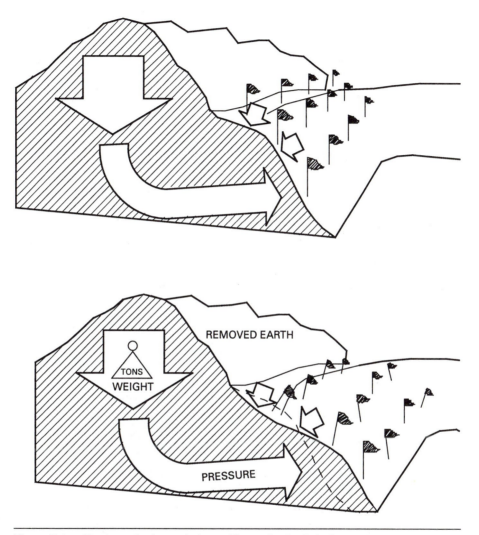

Figure 7-1.   Slope monitoring technique. (Illustration by Dale Barnes.)

the hubs are sinking, the project manager can remove some of the over-burden or have the edges of the excavation shored to prevent a collapse.

In some contracts, excavation work is charged for according to the volume of earth moved. It is often the responsibility of the field engineer to measure the volume of the excavation or the volume of stockpiled material in order to bill or to check billings. This is done by simply measuring the excavation or the stockpile, factoring in the swell or shrink factor of the soil, and computing the volumes using the principles of solid geometry.

It is not unreasonable for the field engineer to request that stockpiles be shaped up so that more-accurate measurements can be taken. This often only takes an hour or two using a loader, dozer, or other earth-moving equipment. Having a more accurate measure may save much more than was spent on the shape-up operation.

Estimating the amount of earth which must be moved on large job sites involves the use of topographic maps, a grid system, and extensive calculations. Many computer programs have been developed which shorten this process considerably. However, due to the rather specialized nature of this area, an in-depth study is beyond the scope of this text. The field engineer is encouraged to consult a good text on the subject of construction estimating to learn the skills of estimating earth moving.

When planning ramps for excavations, the field engineer must know how steep the ramp must be and lay out the ramp area accordingly. This is usually dictated by the available area and the type of equipment that will be using the ramp. If the field engineer is unfamiliar with the ability of various types of heavy equipment to negotiate ramps of varying degrees of slope, the project manager or some other authority should be consulted. If switch-back ramps are to be used, the turning radius of equipment must be taken into consideration.

7-5   Dimension Control of Deep Foundations

After the excavation work has been completed or has progressed far enough for work to safely begin on the floor of the excavation, the field engineer must begin the layout for the foundation components. One of the first decisions that must be made by the field engineer is whether to use surface control points to lay out foundation component locations or establish a control system on the floor of the excavation. Naturally, each method has its advantages and disadvantages.

If electronic instruments are available, giving line and grade information to craftsmen and subcontractors working in the excavation while the instruments are set up on surface points is quick and easy. So much so, in fact, that there may be little or no reason to establish control points and benchmarks in the excavation. This is even more true if working room in the excavation is quite restricted. It may be easier to operate from the surface and stay out of the way of work progressing in the hole. However, this may require the almost constant presence of the field engineer to give line and grade to those working in the excavation.

If the excavation is quite large and extensive work is to be done in the excavation, it may be to the advantage of the field engineer to set up a control system on or near the floor of the excavation. This control system may be used as needed by the craftsmen and subcontractors, with the field engineer having to make only infrequent trips to the site for special layout work or to verify certain dimensions and elevations. In most cases, offset building line control points and elevation benchmarks similar to those used in surface construction must be established in the bottom of the excavation.

Some field engineers use a combination of surface control and excavation floor control systems. It is common to set pier and pile stakes from the surface, then establish a floor control system after the piers or piles are in place. This allows the large pier-drilling and pile-driving equipment to move freely without disturbing control stakes.

Regardless of whether control is maintained from the surface or from the bottom of the excavation, the sequence and layout of points is essentially the same. The sequence progresses naturally from the bottom up. Linear and elevation control is not critical when laying out lower foundation components. The equipment used in the construction of these components is not capable of maintaining a high degree of accuracy. The components are to be covered up and they are designed oversize to allow for some degree of misalignment. Tolerances will be tightened up later when it is time to establish the top of the pier caps and set anchor bolts and plates.

7-6   Layout Procedures for Piers and Piles

Arguments arise as to the definition of a pier as opposed to a pile. For the purposes of this text, a pile is defined as a structural member of wood, concrete, or steel which is driven into the ground. A pier is defined as being made of concrete and being cast in place in a pre-drilled hole.

Piers and piles are usually located at the intersections of building lines, though some may be placed in other strategic locations around the building site. The common practice is to lay out the center of the pier or pile, and stake the location. Hubs are usually not set. The stake is marked as to building line intersection and other information which will tell the operators of the pile-driving or pier-drilling equipment exactly what is required at this location.

The machine operators will roughly center their equipment over the stake and begin drilling the pier holes or driving the piles. Of course, this destroys the stake unless it is removed and kept for reference purposes. The depth and condition of pier holes and the operation of the pile-driving equipment is usually monitored by independent testing laboratories, the architect, the project engineer, or all three.

Diameter, depth, or other design requirements of the piers and piles may be found as notes on the prints or may be contained in a "schedule" similar to a room finish or a window schedule.

The accuracy requirements for laying out the location of the piers and piles are not as strict as for other components of the foundation. The equipment used to place the piers and piles does not lend itself to maintaining extreme accuracy. Also, the design of this type of foundation allows for some latitude in the placement process. If the location stakes are within an inch or two of the exact center line of the pier or pile, this will usually be more than sufficient.

Piles are often driven in clusters. For example, a cluster of six wooden piles may be driven all within a few inches of each other. In this case, the center of the cluster is laid out and the pile-driving subcontractor is left to drive the cluster around the center layout mark.

7-7   Layout Procedures for Pier and Pile Caps

The next part of the foundation to be constructed after the piers and piles are in place are the pier or pile caps. The cap is usually no more than a steel-reinforced concrete cube poured over the top of the piers or piles. The alignment and elevation of the cap is somewhat more critical than the alignment and elevation of the piers and piles.

Pile caps are constructed by one of two methods. One method, shown in Figure 7-2(a), is to excavate around the freshly driven piles and cut them off at the height specified on the prints. Next, reinforcing steel and concrete are placed in the excavation to form the pile cap.

The alternate method, shown in Figure 7-2(b), is to build a form around the protruding ends of the piles and fill the form with concrete to a predetermined level.

Both methods described above require the field engineer to give the elevation of the cut-off point for the piles, the line for the excavation or for setting the forms, and the elevation of the top of the concrete.

Pier caps for cast-in-place piers are constructed much the same way as pile caps. In some cases, a form is constructed above grade and the pile and pile cap are cast in one pour.

In most cases, anchor bolts or reinforcing steel for columns is set in the pile or pier cap. This also requires the field engineer to give line and grade so that the steel or bolts are correctly placed for future construction activities.

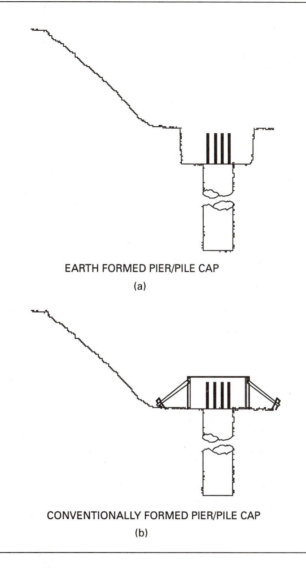

EARTH FORMED PIER/PILE CAP

(a)

CONVENTIONALLY FORMED PIER/PILE CAP

(b)

Figure 7-2.

Layout Procedures for Dowels and Anchor Bolts

Prior to pouring the concrete which will form the pile or pier cap, the method of anchoring the remainder of the building to the foundation must be studied.

If the structure is to have concrete columns, reinforcing steel bars called dowels are set in the concrete of the cap and allowed to protrude upward from the concrete. The reinforcing steel for the columns will be attached to or aligned with the dowel bars and encased in the concrete which will form the columns.

If the structure is to have steel columns, a system of steel anchor bolts and base plates is used to anchor the columns to the cap.

Dowels. The dowels must be placed with some degree of accuracy because the placement of the column reinforcing steel and ultimately the column itself is dependent on proper alignment with the dowels.

The number and geometric design for the placement of the dowels depends on the design of the column. Round columns will usually use dowels arranged in a circle. Dowel placement for square or rectangular columns usually follows the geometry of the column. Square columns have dowels arranged in a square, whereas rectangular columns have dowels arranged in a rectangle.

The dowels are fastened to the reinforcing steel in the pile cap with tie wire. The ironworkers will use the center line of the column location to properly locate the dowels. It is a common practice for the field engineer to place stakes on both building lines on either side of the excavated area where the cap is to be poured or to mark the building lines on the cap forms if the cap is to be formed. The ironworkers can then use string lines to aid in placing the dowels.

NOTE

Some rectangular and square columns have dowels of different sizes in the same dowel cluster.

Anchor bolts. There are some variations in the methods used in anchoring steel columns to the concrete caps. Some methods use base plates which are installed on the job site with the columns being welded to the base plates in the field. Some methods use a base plate which was welded to the column in a fabrication shop and the column and base plate are erected as a unit. In other cases, base plates are set and a column with another smaller base plate pre-attached is set and welded and/or bolted to the first base plate.

Regardless of the base plate-to-column attachment method, the placement of the anchor bolts is critical because the base plates must fit over the anchor bolts. The holes in the base plates are drilled in fabrication shops using equipment which is capable of accuracy to within a few thousandths of an inch. The holes are drilled slightly oversize so that there is some room to adjust the base plate so that it will fit. However, just how much oversize the holes are drilled and how much adjustment is possible varies from job to job. It is not uncommon for 1 " anchor bolt holes to only be drilled 1/16 " oversize. Therefore, it is critical that the anchor bolts be placed accurately not only in relationship to the building line, but also in relationship to each other.

The best way to ensure proper positioning of the anchor bolts is to build a template. Templates are usually built of heavy plywood and supported on the forms or on stakes driven near the excavation where the cap is to be cast.

It is often the responsibility of the field engineer to lay out the bolt pattern on the templates and supervise the construction of the templates. This involves a study of the structural prints of the project to determine the bolt pattern, bolt diameters, and spacing. The field engineer may have to obtain shop drawings from the steel fabricator to determine these dimensions.

CAUTION

Use only **approved** shop drawings. Using unapproved shop drawings may result in incorrect bolt placement.

In most cases, the bolt patterns are laid out with reference to the center lines of the base plate which, in turn, correspond to the building lines. A template cut from plywood should be approximately the same size as the anchor plate but enough larger to allow for supports as required.

After cutting a piece of plywood to the proper size to accommodate the bolt pattern and template supports, lines are scribed on the plywood that represent the building lines.

CAUTION

If the bolt holes are not symmetrical around the building line intersection, the orientation of the template must be **clearly** indicated.

The field engineer may now proceed to lay out the template for drilling. After drilling and installing the support system, it may be advisable to coat the template with concrete form oil. This will preserve the template and ensure that it releases from the dried concrete easily. Also, the exposed bolt threads should be coated with grease. This keeps concrete from sticking to the thread area and aids in the removal of the template after the concrete has set up.

NOTE

In some cases it is possible to install the anchor bolts through the template with washers and nuts holding the bolts in place. The template with the anchor bolts attached may be lowered into the concrete before it sets up. However, if the anchor bolts are quite large, they must be fastened to the steel in the pile cap for support. The template may be placed over the bolts prior to pouring the concrete or lowered over the bolts after the concrete is poured but before it sets up.

CAUTION

Drill observation holes in the template or hold it up off the concrete a few inches in order to observe and ensure that the area under the template is completely filled with concrete and that there are no voids in the bolt area.

When the template is placed in position, the field engineer must determine that (1) it is oriented properly and (2) it is on line in both axes. Once in position, the template must not be disturbed until after the concrete is set.

To monitor the template locations, one or more of three methods may be used. These three methods are described below.

1. **Transit monitoring.** Transits are set up over the building lines that cross the template. The center line of the template is sighted, monitored in both axes, and the template is adjusted as necessary until the concrete is in place and activities have moved on to another area. This requires two transits and two field engineers. One transit, and possibly both transits, must be moved each time concrete placement begins at another cap. This usually does not delay the pour as the transit(s) can be moved much faster than the concrete can be placed. Visibility may be a problem if other activities obscure the view from the control line setup points.

2. **Using a string line from stakes.** Strings are stretched across the cap area from stakes placed on building lines on each side of the cap area. The intersection of the strings corresponds to the intersection of the building lines and the intersection of the lines on the template. One problem with this method is that the string is often some distance above the template and it may be awkward to hold plumb lines against the string and over the template in an attempt to align the template properly. Another problem is that of ensuring that the stakes where the strings are attached have not been disturbed.

This method is similar to using batter boards and is usually accurate enough for setting anchor bolt templates if the difficulties mentioned above can be overcome.

3. **Using a string line from forms.** This method is quite accurate and easily used if the forms are stable and do not move during the pour. A string may be stretched from nails placed on or inside the forms at or near the level of the template. These nails are, of course, set by transits set up on building lines.

Of course, method number one is the preferred method. Regardless of the method used, every effort should be made to have the field engineering crew on hand to monitor the setting of the anchor bolts for any structure.

7-8   Setting Column Base Plates

As mentioned earlier, column base plates may be attached to the column in the field or in the shop. Field attachment means that the column base plate must be set to its correct elevation, leveled, and grouted before the column is brought into place for attachment to the base plate. Grouting is the process of forcing a mortar-like mixture under the base plate. The grout provides a bearing pad for the base plate that is smooth and free of any air pockets. Figure 7-3 shows how a column base plate is mounted on anchor bolts and grouted.

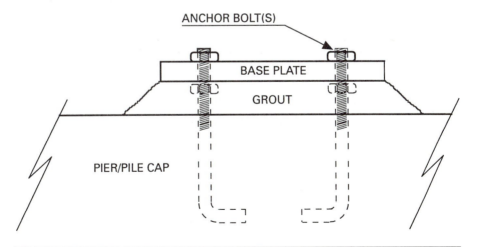

Figure 7-3. Typical base plate mounting method.

In the case of setting the base plate prior to attaching the column to it, nuts are screwed onto the corner anchor bolts. The elevation of the top of these nuts is given by the field engineer by a simple differential leveling procedure. After the nuts are brought to the proper elevation, the base plate is lowered over the bolts and allowed to rest on the corner nuts. At this point, the elevation, alignment, and level condition of the base plate is checked again by the field engineer. The nuts under the plate are used to make adjustments in the elevation and level condition of the plate. This process is similar to leveling a four-screw surveying instrument. Once the base plate is at the proper elevation and is level, additional nuts are placed on the bolts on top of the plate and tightened.

A non-shrinking grout is forced under the base plate, leaving just the corners where the adjusting nuts are located free of grout. The grout is allowed to set up for a few days. After the grout has dried for the specified time, the corner nuts under the anchor plate are backed off until they are clear of the plate. More grout is packed around the corners of the plate where the nuts were backed off to complete the grouting process. At this point, the entire weight of the anchor plate (and whatever is ultimately attached to it) bears on the grout, which in turn bears on the concrete of the cap.

In the case of setting a base plate with the column already attached, the process is somewhat different. The under-plate nuts are set to their predetermined elevation and the base plate with the column attached is lowered over the anchor bolts as above. However, the column is plumbed, instead of the anchor plate being leveled. In plumbing the column, both the under-plate nuts and guy cables are used. Once the column is plumb, the grouting and other processes are the same.

NOTE

On some jobs, leveling and plumbing of columns may be done after a considerable amount of steel has been erected. The grouting may also be done while other work is in progress.

Chapter 7 Review Questions and Exercises

1. How does a slab-on-grade foundation differ from a deep foundation?

2. What extra precautions must be taken when establishing foundation component elevations if an arbitrary finished first-floor elevation of 0.0 ft is specified on prints?

3. What is meant by the "angle of repose" of soil? What effect will this data have on excavation for a deep foundation?

4. Describe techniques by which slope creep and excavation surface subsidence can be monitored.

5. After an excavation for a deep foundation system is open, should the field engineer transfer control points to the floor of the excavation and execute all dimension control from there or should control be maintained from the original surface?

6. Which foundation component requires the more-precise alignment and elevation control, the top of a cast-in-place pier or the top of a pier cap?

7. Which requires the more-precise alignment, dowels or anchor bolts?

8. How does using a template aid in the placement of anchor bolts?

9. List three methods used to monitor the location of anchor bolts and templates.

10. How does leveling a base plate resemble leveling a four-screw instrument?

C H A P T E R

8

DIMENSION CONTROL ABOVE GRADE

8-1 Introduction

The field engineer is responsible for controlling the dimensions of the structure as construction progresses beyond the initial foundation work. The term "getting the job out of the ground" is often used by contractors when referring to these early stages of the project.

This chapter will discuss the responsibilities of the field engineer in providing accurate control lines and grade elevations for the general contractor's personnel as well as the various subcontract trades that may be involved in the project. Of course, in some situations the subcontractor will be required to do his own measurement to ensure the proper location of his equipment, but the field engineer must often provide the basic reference lines and elevations from which other trades must work. He must also occasionally check the work of the subcontractors to ensure that they are in compliance with the plans and specifications in terms of the proper dimension requirements.

8-2 Transferring Construction Control Lines

One of the first tasks the field engineer must perform after the initial basement foundation structure has been completed is that of establishing lines which will indicate where columns will be formed or erected. This involves transferring a set of control lines to the basement slab, or to an area on or near the tops of the piers, pier caps, or footings, and laying out the column lines from these control lines.

Imagine the job site as it must look at this point in the construction process. The excavation is open. There is dirt piled around the perimeter of the hole. Possibly there is a dirt ramp on one side leading down into the hole.

If the building is to have concrete columns, the basement floor slab will probably be in place and there will be small bundles of concrete reinforcing bar sticking up through the concrete slab about four feet into the air. These bundles of four or more bars will be covered with splashed concrete and some will be bent at odd angles. The groups of bars may not appear to be in any particular order or arrangement.

But if one looks closer, the ragged bundles of bars are in lines, though not precisely straight lines! These ragged clumps of reinforcing steel are extending up from the piers or footings below and represent where the columns that support the next floor (and the rest of the building) must be placed. Figure 8-1 shows such a job site.

Figure 8-1.   Typical concrete building under construction. The circle indicates one of several clusters of reinforcing steel where the columns for the next floor will be located. (Photo courtesy of Austin Commercial.)

If the structure is to have steel columns, the basement slab may or may not be in place at this time. If the slab is in place, there will probably be diamond-shaped open areas about three feet square at each column location. Each diamond-shaped opening will have anchor bolts sticking up from the top of the pier, pier cap, or footing below. If the basement slab has not yet been poured, the anchor bolts may have been covered up with dirt during the filling and grading operations in preparation for slab placement. Some hand work may be necessary to expose the bolts and the top of the pier, pier cap, or footing. Figure 8-2 shows such a leave-out in a slab.

It is now the job of the field engineer to establish lines on the newly poured slab from which the carpenters can measure and place the forms for pouring the concrete columns, or to establish lines on or near the piers, pier caps, or footings that the ironworkers can use to properly place steel columns. This is done by establishing control lines based on existing control points and transferring these lines to the construction area.

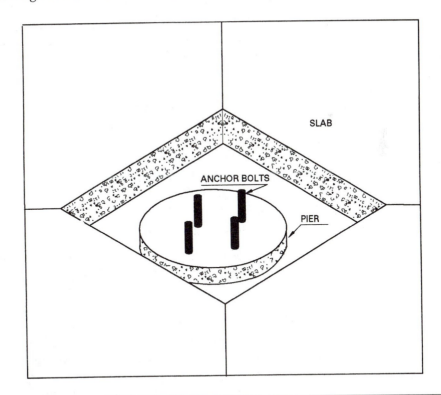

Figure 8-2. Typical slab leave-out for a column.

The placement of control points from which control lines are extended will vary, depending on the shape of the structure, job-site conditions, and local custom. This is discussed in Chapter 6 in some detail. However, for the purposes of clarity and discussion in this section, a variety of control points and monuments will be used as the points from which only two control lines are drawn. The reader should realize that multiple control points surrounding the building site could be used and often are.

8-3   Layout Procedure for Concrete Columns

To begin the process of laying out form lines for concrete columns, the field engineer must first transfer construction control lines to the slab. Once the control lines have been transferred to the slab, the column form lines as well as the location of other items on the slab may be readily found and marked.

The following is a step-by-step procedure for establishing construction control lines and column form lines on a typical basement floor slab. Of course, the same or similar procedures may be followed on each succeeding floor of the structure or for structures with no basement.

Step 1. This operation will require a field engineer and one assistant at the minimum. A transit and a fully stocked field engineer's tool box will be required.

After consulting with the project manager to determine whether any unusual conditions affecting safe operations exist, the field engineer should brief the crew on the operation. Aspects of safety, communications, and procedure should be covered. A set of prints showing the necessary dimensions should be consulted and a copy made from which to work if necessary. The initial field book entries should be made.

Step 2. Move to the area of the excavation and locate a position on the edge of the excavation where the transit may be safely set up. This location must be on a construction control line and must be one from which the field engineer can see the points which establish the control line and at least one edge of the slab. Figure 8-3 shows a typical job site with one N-S construction control line "A" and one E-W construction control line "B". Note the transit positions A and B on their respective control lines.

Set up the transit on the location selected and work it to the control line. Check to ensure that the base of the transit is level and that the lateral movement is locked.

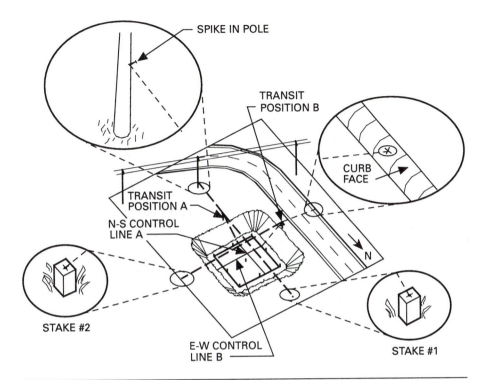

Figure 8-3. A drawing of a typical job site, showing the possible location of control points and control lines.

NOTE

It is a good practice to place a hub stake under the transit at this point. This hub may aid in future sightings.

Step 3. Transfer the first construction control line to the slab. Rotate the barrel of the transit downward and focus on the near edge of the slab. While sighting on the slab edge, direct your assistant to place a reference mark on the slab surface and on the slab reveal on the side nearest the transit. This process has effectively established one point on the control line on the edge and top of the slab at the bottom of the excavation.

Rotate the transit barrel upward, being careful not to disturb the lateral motion of the instrument, and focus on the far edge of the slab. Have your assistant repeat the marking process on the far side of the slab, care-

Figure 8-4.   Method of carrying a control line over a slab edge.

fully using a small square to carry the mark over the edge of the slab if necessary. Figure 8-4 shows this procedure in detail.

Snap a chalk line between the two marks. This establishes the construction control line on the surface of the slab.

 NOTE

If the slab is quite large, intermediate marks should be placed on the surface of the slab about thirty feet apart. This aids in the marking of a straight line completely across the slab. In some cases, the transit must be moved to the slab, worked to the construction line, and intermediate points marked with the transit in the new position. This is particularly true if the slab is irregular in elevation or has large open areas in it.

NOTE

A trick of the trade is to spray the intersections and other strategic locations on the chalk lines with clear lacquer. This preserves the lines much better and enables them to be re-established more easily if necessary.

Step 4. Transfer the second control line to the slab. Move the transit to an adjacent side of the excavation and repeat Steps 2 and 3. This establishes a second construction control line on the slab at 90 degrees to the first one.

NOTE

As noted in Chapter 6, construction control lines may be located along column center lines, at slab edges, on bay center lines, or in some other convenient location that may be dictated by the characteristics of the structure or the job site. In either case, column form lines must be drawn using the construction control lines as reference. This procedure will usually involve offsetting the column form lines from the construction control lines to a preferred location on the newly poured slab so that the form carpenters can begin to erect the column forms.

It is at this point that the field engineer may need to check with the person responsible for setting the column forms (usually the carpenter foreman or some other knowledgeable person) about the most desirable location for the column form lines. The desired placement of this form line may vary from one part of the country to another, but the most common location for the form line is **the edge of the finished column.** This places the line well-clear of the re-bar but does not force the field engineer or the form carpenters to account for form thickness, which may vary considerably. If this method of locating form lines is not the one customarily used in a particular part of the country or if the job dictates another form line locating method, the person in charge of setting the forms should be notified of the method used. Failure to do so may result in the columns being placed in the wrong location.

Step 5. Locate and mark the column form lines using the construction control lines for reference. The column form lines will be parallel to the construction control lines. The process of establishing these parallel lines is known as offsetting. To offset a column form line from a construction control line, several methods can be used depending on the degree of accuracy required and the design of the structure. A combination of methods may be necessary, and one method can be used to cross check another for accuracy.

Method 1. Measure along the slab reveal.

A. Measure the distance from the construction control line to the column form line along one slab reveal at 90 degrees to the construction control line. Mark the column line termination point on the slab. See Figure 8-5.

B. Move to the opposite side of the slab, repeating the measurement and marking, and snap a chalk line between the two points.

This method is acceptable if the layout of the slab lends itself to using the slab edges for reference lines, and if the specification tolerances will allow the level of accuracy this method produces.

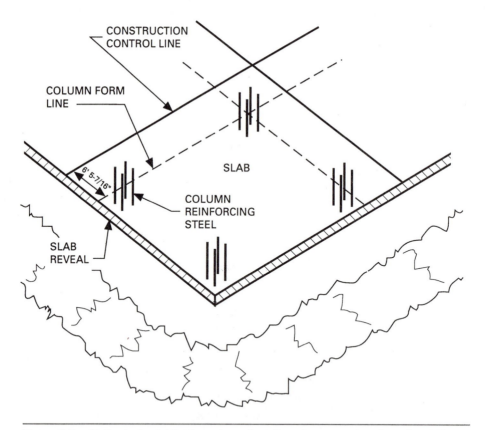

Figure 8-5. Method of establishing column form lines by measuring along the slab reveal.

Method 2. Carpenter's square method.

A. Place a carpenter's square or other square item (a half sheet of plywood will do if the mill edges are used) with one arm or edge along the construction control line chalk line previously marked on the slab. See Figure 8-6.

B. Stretch a chalk line across the slab surface and adjust it until it is parallel to the 90-degree arm of the square or edge of the plywood. Snap the line.

This method is not as accurate as others, but may be sufficient in buildings that are to be only one or two stories tall and buildings in which precise column placement is not critical.

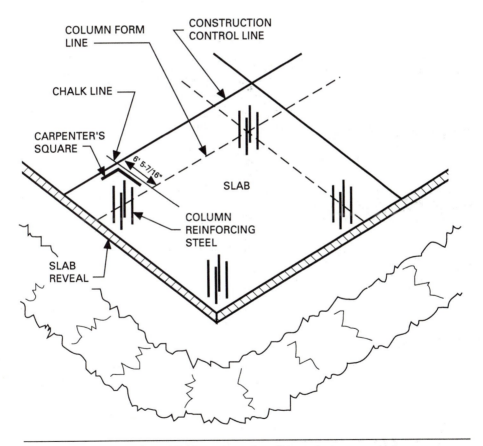

Figure 8-6. Method of establishing column form lines using a carpenter's square.

Method 3. Trigonometry method.

A. Choose an unobstructed location on the construction control line where a parallel column form line can be laid out.

B. Lay out two 30-60-90 triangles twenty to thirty feet apart with the "a" side of the triangles on the construction control line and the "b" lines pointing in the same direction but at 90 degrees to the construction control line. See Figure 8-7.

The length of the "b" line of each triangle should be the same as the distance from the construction control line to the column form

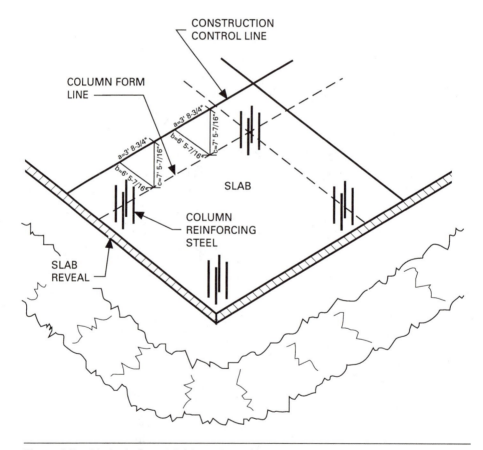

Figure 8-7. Method of establishing column form lines using trigonometry.

line. Using the formula a = b tan 30, derive the length of the "a" line. Using the formula $a^2 + b^2 = c^2$, derive the length of the "c" line.

C. Once the two triangles are laid out on the slab, stretch a chalk line across the "b" and "c" line intersection points of each triangle and snap it. This line is the column form line and is parallel to the construction control line. It may be extended as necessary by chalk line, string line, or by working the transit to the line and sighting along the line.

Step 6. Lay out intersecting control lines.

A. Move the transit to an adjacent side of the excavation and repeat steps 2 through 5.

NOTE

From this point, the same procedure is repeated until all necessary column form lines have been marked on the slab. It is common on most projects to lay out only two column form lines per column, one at 90 degrees to the other. It is assumed that the carpenters can work from one line on one side of the column and one line on an adjacent side. This is particularly true if the columns are square.

Step 7. Inspect the work.

A. Move to a high vantage point. Observe the work. Check to see if the lines are parallel and if the proper lines have been laid out to reflect different sizes of columns if necessary.

B. Check the prints. Do you have the right page? Spot check the dimensions for accuracy.

C. Use a square to check for 90-degree corners where necessary.

Step 8. Record the procedure, dimensions, and locations in a field notebook.

A. Record the location of the transit and which control points were used.

B. Record the method used in establishing the column form lines.

C. Record the time of day, elapsed time, weather, and the names of the personnel involved in the procedure.

D. Record the page and date of the latest revision of the blueprints used.

E. Record any unusual events which may have occurred prior to, during, or after this procedure which may have an effect on the results.

Step 9. Secure the equipment, field notes, and prints in accordance with accepted procedures. Report that the operation has been completed to the proper authority.

CAUTION

Caution should be used in reading the prints for column form line layout. It is common for concrete columns to be rectangular, not square. This is particularly true in some parking garages where parking space is at a premium and in structures where the columns are to be used for shear walls.

Columns may not be located in the same position on all floors. There may be transfer beams involved which will vary the lateral location of columns on adjacent floors.

Column sizes may vary from floor to floor, usually decreasing in size as the building gets taller. Check the schedules and the structural prints for column sizes on each floor. Only approved structural (not architectural) prints should be used.

After the columns are formed and poured, construction of the next floor can begin. This involves setting forms on which concrete is poured or setting beams on which decking is fastened. Concrete may be poured on the decking.

In either case, there are usually points on the top of the floor form where no concrete should be placed. These areas are called leave-outs. This includes openings for stairs, elevators, utility lines, and so forth. It is often the responsibility of the field engineer to mark the locations for these leave-outs.

To accomplish this, the control lines must be transferred to the forms. This is done in the same manner as transferring the control line to the slab. Once the control lines are marked on the forms, the leave-out areas may easily be laid out.

NOTE

The field engineer should lay out the leave-out areas before reinforcing steel is placed for the floor slab. It is also a good idea to get any chalk lines down on the floor forms before the form release oil is sprayed on the forms. Chalk lines do not work well on oily forms.

8-4   Layout Procedure for Steel Columns

There are two general methods used to mount steel columns to concrete foundations: the prefabricated column method and the field assembly method. Each method has its advantages and disadvantages and each is more appropriate in certain conditions than the other.

1. **Prefabricated column method.** The columns are fabricated off site in a steel fabricator's shop. The columns are shipped to the job site with the base plates and other fastening tabs installed and with all necessary holes drilled. If the column is to be mounted on anchor bolts, the base plates will have anchor bolt holes pre-drilled to the specified pattern. Often, if the anchor bolts have been carefully set with templates, there is no need for further alignment of the columns. They are simply erected and set over the anchor bolts. The nuts and washers are installed, the columns are plumbed, and the anchor nuts torqued. Smaller, lighter prefabricated columns, those made of pipe or round or square structural tubing, are sometimes called "lally" columns.

2. **Field assembly on pre-installed base plate method.** If the columns do not have the base plates installed when delivered to the job site, they will usually be set on pre-installed base plates and field welded in place. These pre-installed base plates may be mounted on anchor bolts or they may have bolt-like protrusions called studs welded to the backside of the plate and embedded in the concrete of the foundation. In either case, the column must be properly positioned on the base plate for welding.

Regardless of the column connection method specified on the plans and in the specifications, it is the responsibility of the field engineer to provide reference lines that the ironworkers can use to properly place the columns when necessary.

The following is a step-by-step procedure for establishing layout lines for steel columns.

Step 1. Same as for concrete column layout as described above.

Step 2. Same as for concrete column layout as described above.

Step 3. Transfer the first construction control line to the area near the column mounting points. Rotate the barrel of the transit downward and focus on a point close to the foundation on the near side of the structure.

Direct your assistant to drive a stake so that the top of the stake is visible and the cross hairs of the transit scope fall on the top of the stake. Place a mark or tack on the top of the stake at the point indicated by the cross hairs of the transit. Rotate the transit barrel upward and repeat the process, setting a stake with a tack or mark on the top on the opposite side of the structure.

This has established two points on the construction control line on the top of two stakes, one on either side of the structure. From this line, it is possible to lay out other lines and locate other items that must be placed in the structure.

Step 4. Transfer the second construction control line to the area near the column mounting points. Move the transit to an adjacent side of the excavation and repeat steps 2 and 3. This has established four stakes with tacks or marks on their tops that correspond to two construction control lines, one at 90 degrees to the other.

NOTE

At this point, the field engineer must make a decision as to whether the column erection lines will be placed directly on the center axis of the column lines or whether the column erection lines will be offset to one side of the column center lines. This will depend upon the type of column mounting method specified as well as other factors.

The following rules will aid the field engineer in deciding which location is best for the column erection lines in a given situation.

Rules for locating steel column erection lines <u>on</u> the column center lines.

1.   There must be a clean, flat surface on which the column erection lines can be scribed. The area should be larger than the column base plate by at least ½ inch.

2.   The column base plates should be installed at a fabrication facility, rather than field welded.

3.   The column base plates should be of uniform size.

4.   The column base plates should be welded to the columns so that the axes of the base plates conform to the axes of the column. That is, the column should be precisely centered on the base plate.

5. Anchor bolts must be set precisely using templates and not disturbed during concrete placement.

6. Any anchor bolt holes in the column base plates should be precisely drilled to conform to the anchor bolt template used to set the anchor bolts, and vice versa.

Rules for locating steel column erection lines _alongside_ the column center lines.

1. If any of the rules listed above are violated, the field engineer should consider placing the column erection lines alongside the column center lines.

2. Use this technique if the columns are to be field welded to pre-installed base plates or plates embedded in the foundation structures or the slab. The embedded plates may have shifted during concrete placement, therefore the ironworkers may not be able to place the column in the exact center of the plate and still have the column in proper alignment.

3. If the anchor bolts are out of position, requiring the use of a torch to elongate column base plate holes. (The torching of column base plate holes must usually be approved by an engineer.)

4. If several different sizes of columns are to be installed on one column line or if any of the AISC W or S shape members (those with a cross section resembling an "H" or "I") columns are used and some are oriented with the flanges on opposite axes. That is, the end columns may have the flanges parallel to the N-S lines while interior columns have the flanges parallel to the E-W lines.

5. If there are obstructions that will prevent viewing or marking the column axes locations.

6. If the transit used to locate the bases of the columns is to be used to plumb the columns. It is easier to measure from a line of sight over to a column face during the plumbing operation. The same thing applies if a laser is to be used for plumbing a line of columns.

Step 5. Establish offset lines or center lines for column placement. If the choice has been made to use a column erection line which is to run alongside the column center lines, two options are open to the field engineer. One option is to run a string line from stakes placed on the offset line. The ironworkers then simply measure from the string to the face of the column, compensating for one-half the thickness of the column at the point of tape contact.

The other option is to set up a transit on the offset line, sighting down the line alongside the columns. The ironworkers then measure from the line of sight over to the face of the column, again compensating for one-half of the column thickness at the point of tape contact.

Care must be taken in each case to ensure that the tape is level when measuring from the line of sight or the string. Of course, a laser level set up on the offset line and projecting along the line in the vertical plane could be used as well.

In either case, it is good practice for the field engineer to communicate the offset distance to the ironworkers who will be setting the columns, particularly if the field engineer is not going to be present when the actual erection of the columns takes place. There is no recognized national standard as to how much an offset column erection line should be offset from the column center lines.

If the choice has been made to run the column erection line on the column center line, set the transit up **on** the column center line, sighting down the line toward the position of the last column on the line. Direct your assistant to scribe a line on each column mounting point on the line of columns.

Continue the procedure until each column mounting point has been marked with a set of lines which should cross at the exact center of each column location.

Step 6. Inspect the work.

A. Move to a high vantage point. Observe the work. Do the lines appear to be in the right place relative to the foundation work? Do the lines run through or parallel to anchor bolt clusters? Are there any anchor bolt clusters not covered by a set of column erection lines?

B. Check the prints. Do you have the right page and the correct revision if there is one? Spot check dimensions for accuracy.

C. Use a square to check for 90-degree corners where appropriate.

D. Are stakes straight and marked properly? Are the guard stakes flagged?

Step 7. Record the procedure, dimensions, and locations in a field notebook.

A. Record the location of the transit and which control points were used.

B. Record the method used in establishing the offset lines.

C. Note whether the column erection lines were run on the column lines or if they were offset. If offset, record the offset distance.

D. Record the time of day, elapsed time, weather, and the names of the personnel involved in the procedure.

E. Record the page and date of the latest revision of the blueprint used.

F. Record any unusual events which may have occurred prior to, during, or after this procedure that may have an effect on the results.

Step 8. Secure the equipment, field notes, and prints in accordance with accepted procedures. Report that the operation has been completed to the proper authority.

NOTE

Often, when erecting columns with pre-installed base plates, the column base plate bolts are not torqued and the columns are not plumbed during the initial erection process. The idea is to get the steel erected quickly and release the crane, trucks, and some of the ironworkers to do other work. The torquing and plumbing is done after erection, when the field engineer and the ironworkers can devote the time to measuring, setting up offset lines, and plumbing the columns. The plumbing of the columns can usually be done at the same time that the column bases are tapped into final position with hammers.

It must be stressed again that the positioning of anchor bolts is critical, as column base plate bolt holes are not usually drilled more than a few fractions of an inch oversize. This allows for little lateral movement of the column bases to bring them into line. If the base plates do not fit the anchor bolts precisely, time is wasted while the ironworkers enlarge the bolt holes with cutting torches.

8-5   Layout Procedure for Columns on an Arc

Many modern buildings are built in the shape of curves or arcs rather than rectangles. This practice presents some challenges to the field engineer in giving line and grade for dimension control during construction. Larger projects may have one benchmark or project control point at the center or radius point of a pie-shaped job site. Other projects may have several control points. Locating columns and other structures on a site with curved lines can usually be done with a system of coordinates, by using an electronic distance measuring instrument (EDMI) to accurately measure angles and distances from the central control point, or by using the curve layout method described in Chapter 10.

The following is a step-by-step approach to laying out columns on an arc where a single control point is used and the point is accessible.

Step 1. This operation will require a field engineer and one assistant at the minimum. A transit and a fully stocked field engineer's tool box will be required.

After consulting with the project manager to determine whether any unusual conditions affecting safe operations exist, the field engineer should brief the crew on the operation. Aspects of safety, communications, and procedure should be covered. A set of prints showing the necessary dimensions should be consulted and a copy made from which to work if necessary. The initial field book entries should be made.

Step 2. Set up the transit over the master control point. Work the transit to the project master control point and focus on the second reference point on the master construction control line.

Step 3. Establish the column center locations on the arc lines. Beginning at the master control point, measure along the master control line a distance equal to the length of the radius of the front column arc line. Mark this point X. Do the same for the rear column arc line.

From this point, much depends on how much information is available to the field engineer. In some cases, the architects and engineers provide extensive measurements relating to the arc layout. In other cases, as little information as the angle between column lines and the radius of the arc is all that will be shown. It is the responsibility of the field engineer to calculate the necessary missing dimensions in order to complete the layout.

The least difficult and the most accurate way to lay out the column locations on arc lines is to determine the length of the chord between the column centers and lay out the column locations from the end of one chord line to another. The following are the procedures and steps to accomplish this type of column center-to-center layout.

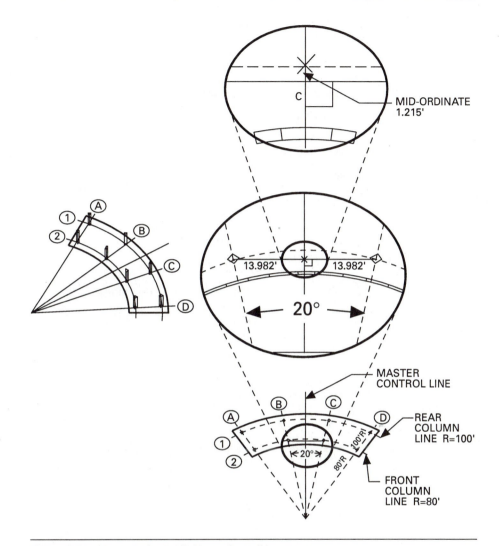

Figure 8-8.   Method of establishing column lines on an arc.

A. Refer to Figure 8-8. Measuring along the master control line to the column arc line, the chord line between columns will cross the master control line some distance short of the arc line. This distance "M" (for mid-ordinate) must be determined and deducted from the radius to determine the point where the chord line crosses the master control line.

To determine "M," use the following formula:

$$M = R - \left[R\left(\cos \tfrac{1}{2}\,\text{delta}\right)\right]$$

In the example shown in Figure 8-8, the following data are given:

R = 80 ft (front column line)
R = 100 ft (rear column line)
delta = 20 degrees

Determine: cos ½ delta = cos 10 degrees = .98481

Working the above data into the formula:

$$M = R - \left[R\left(\cos \tfrac{1}{2}\,\text{delta}\right)\right]$$
$$= 80\ \text{ft} - [80\ \text{ft} \times .98481]$$
$$= 80\ \text{ft} - 78.7848\ \text{ft}$$
$$M = 1.215\ \text{ft (rounded to three decimal places)}$$

This value corresponds to the "a" side of a right triangle. **Note:** Repeat the process for the rear column line using the 100 ft R value.

B. Measure back along the master control line toward the transit 1.215 ft from the "X" previously marked and mark point "C." This point is the right angle "C" of a right triangle formed by lines extending from the master control point to the column center and back to "C."

C. To determine the chord length "Lc" from one column center to the nearest adjacent column center, use the formula:

$$Lc = 2R\ (\sin \tfrac{1}{2}\,\text{delta})$$

Given: delta = 20 degrees

Determine: ½ delta = 10 degrees and sin 10 degrees = .17365

Working the above data into the formula:

$$Lc = 2R\ (\sin \tfrac{1}{2}\,\text{delta})$$
$$= 160\ \text{ft} \times .17365$$
$$Lc = 27.784\ \text{ft}$$

As the master control line was previously set to bisect the delta angle, it must also strike the chord at mid-point. Thus the distance from the master control line to the column center is half of Lc or 13.892 ft. This value corresponds to the "b" value in the right triangle described above.

D. Using the formula $a^2 + b^2 = c^2$, solve for c. (Given: a = 1.215 ft and b = 13.892 ft)

Working the above data into the formula:

$$a^2 + b^2 = c^2$$
$$1.4767 + 192.9877 = 194.4644$$
$$c = \sqrt{194.4644}$$
$$c = 13.945 \text{ ft}$$

This value corresponds to the "c" side or hypotenuse of the right triangle.

E. Lay out a right triangle using the dimensions and the procedure described earlier in this chapter. The intersection of lines b and c will be the center of the column. Sight the transit on the newly located column center and mark a line through the center from front to rear. This line will locate the opposite corners of the column or will locate the mid-point of opposite sides of the column, depending on the orientation of the column.

Step 4. Completing the process. Repeat the process of locating the column centers as necessary. It may be possible to use previously located column center lines for control lines as the process continues.

To find the rear column centers, simply extend the line from the transit through the center of the front columns previously located, toward the rear of the structure. Measure along this line a distance equal to the difference between the radius of the front column line and the rear column line, marking the column centers accordingly.

Step 5. Repeat steps 6 through 8 of Section 8-4.

8-6   Leave-outs for Plumb Lines, Lasers, and Other Measuring Devices

Often it is possible to provide an opening in each successive floor of a structure so that the openings are aligned vertically. This allows for a laser beam to be projected upward through the building, giving a point from which measurements can be taken. In some cases, a plumb bob may be lowered through the opening and measurement taken from the plumb line. If the plumb line must be carried through several floors, stainless wire is often required and the plumb bob may weigh several pounds. The plumb bob weight may have to be suspended in a bucket of heavy oil to dampen the swing. The plumb line may even have to be run through large conduit with "windows" cut in the conduit at each floor from which measurements may be taken.

Levels and transits with optical plumb bobs can be used over leave-out openings, but the distance of their downward focus is usually limited. On relatively short structures, plumb lines and optical plumbing can be done through leave-outs with the leave-outs being covered as the lines are carried to the next floor. The following is the procedure used in this case.

A. Center the transit over the opening and over the reference mark on the floor below. Use the optical plummet or a string bob as necessary.

B. Once the instrument is centered over the reference mark on the floor below, retract the string line. Being careful not to disturb the instrument, place a piece of plywood over the hole and fasten it into place with concrete nails, powder-actuated tool pins, or some other semi-permanent method.

C. Lower the plumb bob to a point near the surface of the plywood or sight through the optical plummet and mark the location on the plywood.

D. Move the instrument up a floor and center it over the mark placed in C above.

This process has effectively transferred a control point up one floor. The work may progress on the floors below without interruption and the leave-outs below may be closed.

If an upward-projecting laser is used to carry a line up through the structure, similar methods may be used. Of course, it is best to leave the laser in position for as long as possible, projecting up through several floors. But if work on the lower floors requires that the laser be moved, the following procedure may be used.

A. Place a piece of white material (or other material that will show the laser beam) over the leave-out hole **one floor above** the floor where the laser is to be relocated. Mark the location of the laser beam on the **bottom** of the material covering the hole.

B. See that the material is not disturbed and move the laser to the new location on the floor below the material. This may involve having to cover the leave-out hole where the laser is to be set up, with plywood fastened to the floor in a semi-permanent manner as previously described.

C. Set up and level the laser at the new location, ensuring that it is projecting upward and that the beam is hitting the mark previously placed on the material above. Once the field engineer is satisfied that the laser is level and secure, and that it is hitting the mark on the material above, the material may be removed and work may continue.

8-7   The Four-foot Mark

After the columns are in place, the field engineer may be required to give grade (elevation indications) to aid other trades in installing their equipment or materials. For example, the air conditioning subcontractor must set his ducts at a certain height to clear the ceiling grid and possibly some fire sprinkler plumbing lines. This means that he must have a benchmark from which he can measure while installing the duct work.

The reference point for each floor is customarily set at four feet above the finished floor level of that floor. This four-foot mark is transferred in from the job-site benchmark and indicated on columns, walls, or other convenient places where it is readily accessible to the various trades that will need to use it as a reference. The indicator marks resemble a transit and will usually have the "4' AFF" indication nearby. See Figure 8-9. This, of course, means that the line

Figure 8-9. A typical mark indicating a point four feet above finished floor.

extending through the mark representing the barrel of the transit is "four feet above finished floor." In some instances, the field engineer will also indicate the true elevation of the symbol based on the job-site benchmark. Additional information on transferring benchmark elevations can be found in Chapter 2.

8-8 High-rise Construction Practices

In some high-rise construction projects the elevation measurements must be so carefully controlled that special efforts must be taken to ensure the highest accuracy in defining vertical dimensions. For example, a surveyor's tape hung in an elevator shaft or other vertical opening cannot just be hung, aligned with a mark, and read. The standard techniques of high-accuracy horizontal linear measurement must be followed even when the tape is in the vertical position. As described in Chapter 4, proper techniques for taping, such as calibration, temperature correction, tensioning, and so forth, must be followed. However, the weight of the tape should be deducted from the horizontal tension forces normally required because the tape is hanging in the vertical position. This may be further complicated if the entire length of the tape is not free to hang in the vertical position. In this case, the proportion of the weight of the tape left to freely hang vertically is deducted from the tension force recommended.

High-rise buildings will "shrink" in height as construction progresses. This is due to the shear weight of the upper parts of the structure pressing down on the floors below and, possibly, to foundation subsidence. In some cases, the architects and engineers will make allowances for this shrinkage in their designs. In other cases, the shrinkage will have no effect on structural components and is ignored. The field engineer must be aware of this phenomenon and should consult with the architects and engineers to determine the best method of establishing vertical control before the job progresses beyond the foundation construction phase.

In some cases, the elevations of the upper floors will be established by measuring upward from a ground-level benchmark established outside the building. In other cases, the elevations of upper floors will be established by simply measuring from the floor immediately below.

Plumbing high-rise buildings can present a challenge to the field engineer. High-rise buildings sway in the wind. The field engineer must be prepared to recognize this sway and to take the average or center-of-swing readings as necessary.

The practice of placing lasers pointed upward through leave-outs as described earlier is common, but lasers and transits may be erected on temporary platforms protruding from the corners or sides of buildings as well. These platform positions require that a field engineer who is not afraid of heights be employed and that some weather protection be available for the personnel as well as the instruments.

Some large structures and high-rise buildings may require that the field engineering work be done at night or very early in the morning. This is due to several factors.

1. The sun may cause one side of the structure to expand more than the other side.
2. The sun may cause heat waves, preventing accurate sighting with transits. These heat waves may affect lasers and EDMIs as well. It is a good practice to shade the instrument in these conditions.
3. Linear measurements over long distances may not be accurate if the tapes are used in intense sun and shade on the same line.
4. Work in progress and dust may prevent sighting during the day.
5. In some locations, winds die down at night or near dawn, and sightings must be taken at those times.

Often, field engineers will try to use adjacent buildings for sighting platforms during high-rise construction. This practice should be approached with caution for several reasons.

Tall buildings sway. If the platform building is swaying and the building under construction is swaying, the field engineer may not be able to get a neutral reading on certain dimensions.

It is not uncommon for the owners or tenants of adjacent buildings to initially agree to the use of their building as a sighting platform only to change their minds as the project progresses. This may leave the field engineer without a critical benchmark, instrument location point, or control point.

Chapter 8 Review Questions and Exercises

1. What does the phrase "getting the job out of the ground" mean?
2. When establishing column form lines on a slab, is the recommended location the edge of the finished column or the outside edge of the form?
3. List the three methods of laying out offset or parallel lines.
4. Is the thickness of the web of a steel column of any concern when laying out offset lines for steel column erection?

5. When steel columns are being plumbed, does the field engineer need to be on the site?

6. When should the field engineer have an ironworker use a cutting torch to enlarge bolt holes in column base plates?

7. When locating columns on an arc, does the field engineer need to have access to the arc center point?

8. What is the significance of the "four-foot mark"?

9. List several problems associated with dimension control in high-rise construction that are not normally encountered when building low-rise structures.

10. What are the advantages of doing field engineering work on high-rise structures at night?

CHAPTER

9

TOPOGRAPHIC MAPS: THEIR PRODUCTION AND USE

9-1   Introduction

To describe the topography of an area is to describe the general condition of the earth's surface in that area. The area could be described as being flat or hilly, and containing streams, lakes, mountains, and man-made objects. The elevations of certain points in the area would normally be included in the description as well. These elevations could be keyed to an arbitrary point for this particular area or they could be keyed to sea level datum. (See Section 2-1.)

When planning a construction project, the topographic description of the property where the project is to be built must be presented in a graphic format which can be easily interpreted by those who have an interest in the property and the proposed project. This graphic presentation is most often in the form of a topographic (topo) map. A topo map will show the location and elevation of all pertinent natural features and an outline of any man-made structures located on the property. The elevations of prominent points on the man-made structures are usually taken at the top of the foundations, slabs, or other easily observed points at or near grade level.

It is often the responsibility of the field engineer to gather the data from which some topo maps are drawn. Additionally, the field engineer may have to interpret topo maps that have been prepared by others. Interpretation may include such activities as producing estimates of the volume of earth which will have be moved during construction, locating certain features for demolition or modification, and producing profile studies for engineering or planning purposes.

9-2   Map Standards

In construction activities such as engineering, estimating, and building, graphic representations (blueprints) of the structure are used daily. These graphics are prepared according to a common set of standards. Because all drafters use these common standards, anyone who is familiar with these standards can "read" and understand the blueprints prepared by others.

The same is true for topo map drafting. Certain standards are used to depict various features or conditions existing on the surface of the earth in the area under study. Due to the complexity of map drafting, the student should consult a reputable text on the subject if a detailed and in-depth study is required. Only some of the most basic standards and procedures will be explored in this chapter.

Scale

The scale of the topo map, as with any map, is dependent on several factors. Perhaps the most important factor to be considered is the detail required for the map to do its intended job. Of course, the greater the detail required, the smaller the area that can be covered.

For example, many maps are prepared in a scale of 1:24,000. This means that one inch on the map represents 24,000 inches (approximately $\frac{1}{3}$ mile). At this scale, it is impossible to draw much detail on the map.

Topo maps used in construction usually cover the project site and a short distance beyond. If the site is quite large, a map may be found in the prints which covers the entire area. This large scale map will be accompanied by several pages of maps drawn at different scales which cover smaller portions of the job site.

Some companies try to standardize their map-drafting activities so that only certain size scales are used. This makes it easier for planners and estimators to visualize and calculate using these maps. Of course, with the growing use of computers and plotters, maps can be drawn and plotted to almost any scale desired.

Symbols

The symbols used in topo map drafting are usually self-explanatory or are explained in a legend shown on the map or on an adjacent page. Figure 9-1 shows many of the most frequently used symbols. Remember, the symbols

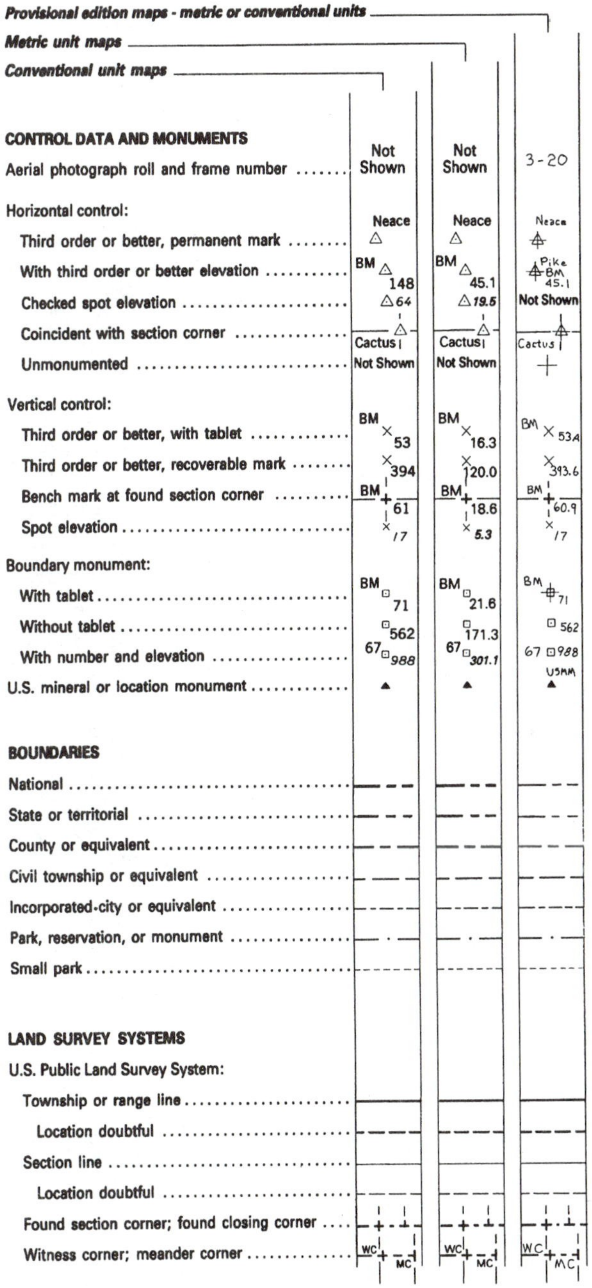

Provisional edition maps - metric or conventional units _____

Metric unit maps _____

Conventional unit maps _____

	Conventional unit maps	Metric unit maps	Provisional edition maps
CONTROL DATA AND MONUMENTS			
Aerial photograph roll and frame number	Not Shown	Not Shown	3-20
Horizontal control:			
Third order or better, permanent mark	Neace △	Neace △	Neace ⊕
With third order or better elevation	BM △ 148	BM △ 45.1	⊕ Pike BM 45.1
Checked spot elevation	△ 64	△ 19.5	Not Shown
Coincident with section corner	△ Cactus	△ Cactus	⊕ Cactus
Unmonumented	Not Shown	Not Shown	+
Vertical control:			
Third order or better, with tablet	BM × 53	BM × 16.3	BM × 53A
Third order or better, recoverable mark	× 394	× 120.0	× 393.6
Bench mark at found section corner	BM + 61	BM + 18.6	BM + 60.9
Spot elevation	× 17	× 5.3	× 17
Boundary monument:			
With tablet................................	BM □ 71	BM □ 21.6	BM ⊞ 71
Without tablet	□ 562	□ 171.3	□ 562
With number and elevation	67 □ 988	67 □ 301.1	67 □ 988
U.S. mineral or location monument	▲	▲	USMM ▲
BOUNDARIES			
National	— — —	— — —	— — —
State or territorial	— —	— —	— —
County or equivalent.......................	— —	— —	— —
Civil township or equivalent	— —	— —	— —
Incorporated-city or equivalent	— —	— —	— —
Park, reservation, or monument	— —	— —	— —
Small park.................................			
LAND SURVEY SYSTEMS			
U.S. Public Land Survey System:			
Township or range line			
Location doubtful	— — —	— — —	— — —
Section line			
Location doubtful	— — —	— — —	— — —
Found section corner; found closing corner	+ +	+ +	+ +
Witness corner; meander corner..............	WC + MC	WC + MC	WC + MC

Figure 9-1. Topographic map symbols. (Courtesy of U.S. Department of the Interior.)

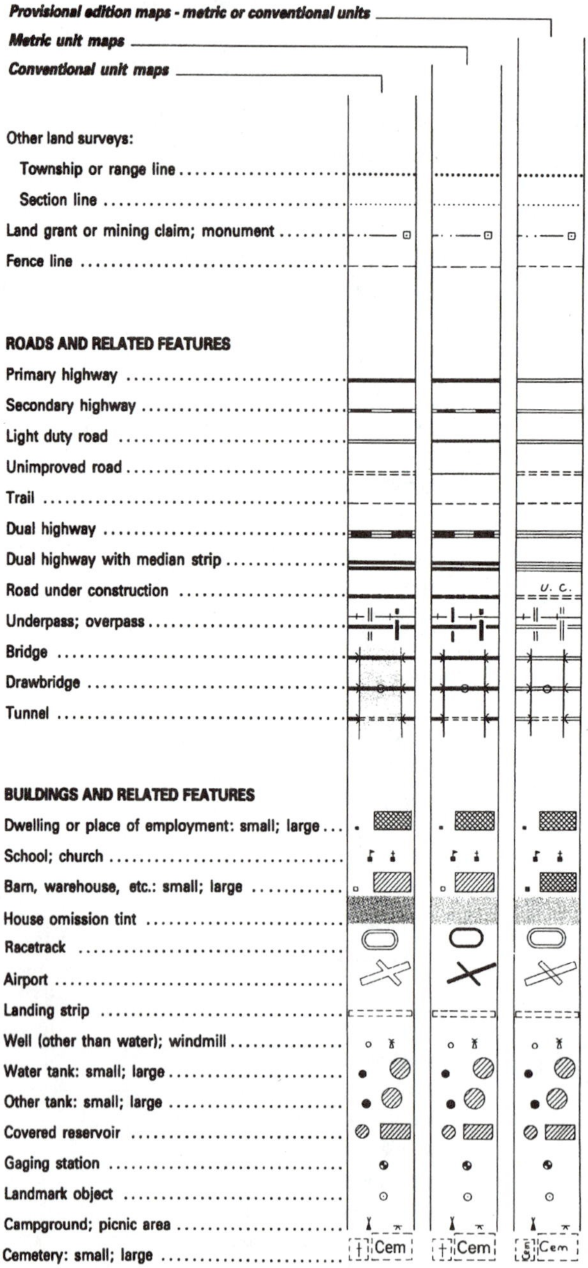

Provisional edition maps - metric or conventional units

Metric unit maps

Conventional unit maps

Other land surveys:

Township or range line

Section line

Land grant or mining claim; monument

Fence line

ROADS AND RELATED FEATURES

Primary highway

Secondary highway

Light duty road

Unimproved road

Trail

Dual highway

Dual highway with median strip

Road under construction

Underpass; overpass

Bridge

Drawbridge

Tunnel

BUILDINGS AND RELATED FEATURES

Dwelling or place of employment: small; large

School; church

Barn, warehouse, etc.: small; large

House omission tint

Racetrack

Airport

Landing strip

Well (other than water); windmill

Water tank: small; large

Other tank: small; large

Covered reservoir

Gaging station

Landmark object

Campground; picnic area

Cemetery: small; large

Figure 9-1. *(Continued)*

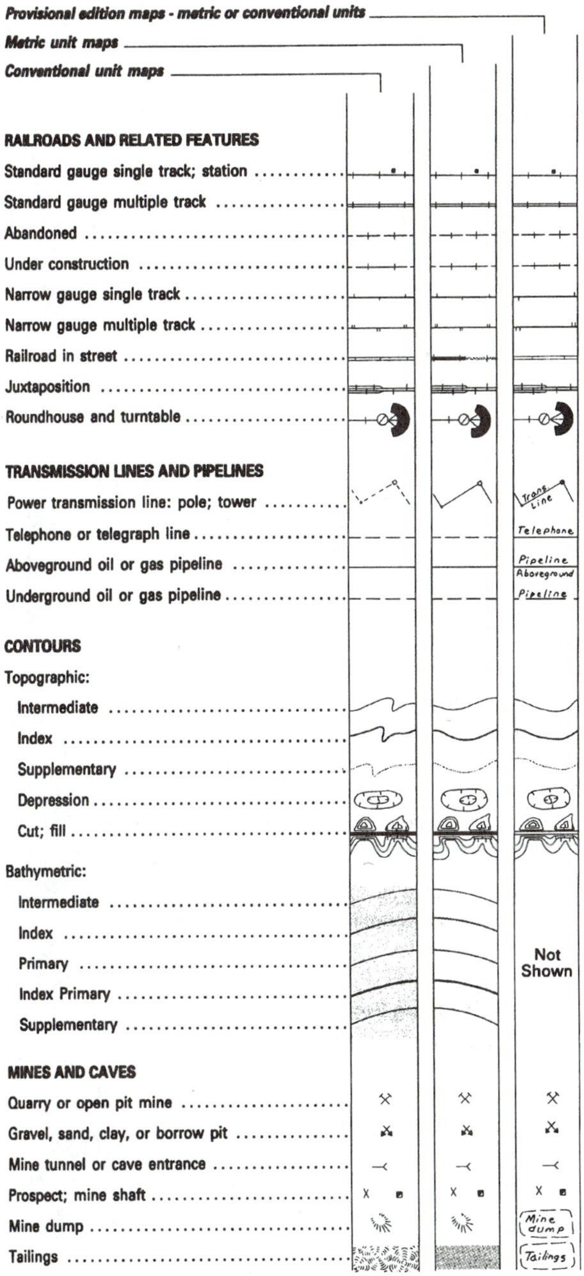
Provisional edition maps - metric or conventional units
Metric unit maps
Conventional unit maps
RAILROADS AND RELATED FEATURES
Standard gauge single track; station
Standard gauge multiple track
Abandoned
Under construction
Narrow gauge single track
Narrow gauge multiple track
Railroad in street
Juxtaposition
Roundhouse and turntable
TRANSMISSION LINES AND PIPELINES
Power transmission line: pole; tower
Telephone or telegraph line
Aboveground oil or gas pipeline
Underground oil or gas pipeline
Trans. Line
Telephone
Pipeline
Aboveground
Pipeline
CONTOURS
Topographic:
Intermediate
Index
Supplementary
Depression
Cut; fill
Bathymetric:
Intermediate
Index
Primary
Index Primary
Supplementary
Not Shown
MINES AND CAVES
Quarry or open pit mine
Gravel, sand, clay, or borrow pit
Mine tunnel or cave entrance
Prospect; mine shaft
Mine dump
Tailings
Mine dump
Tailings

Figure 9-1. *(Continued)*

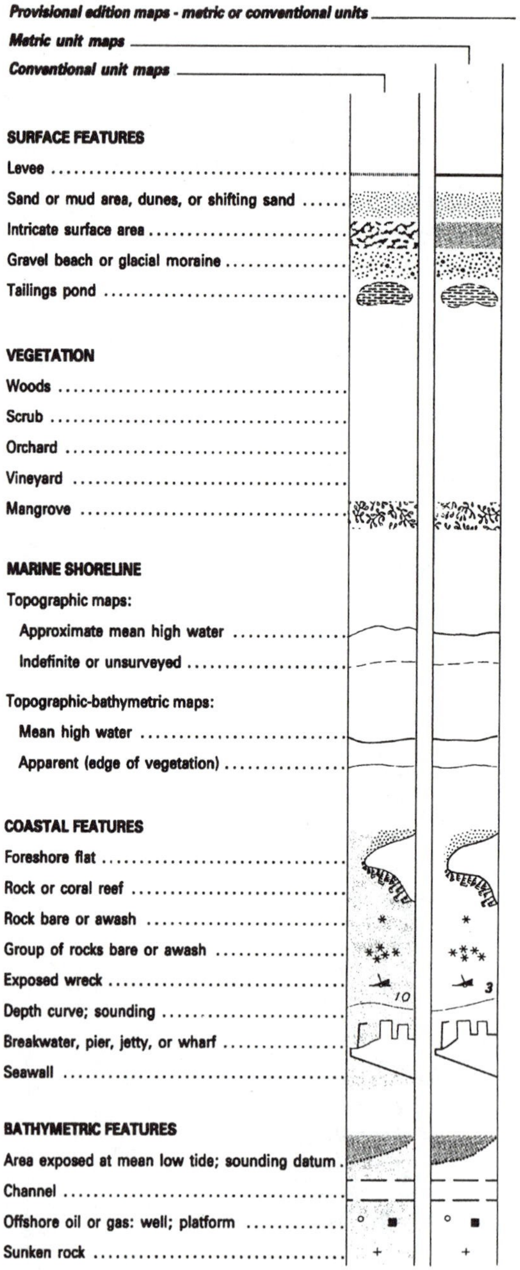

Provisional edition maps - metric or conventional units _____

Metric unit maps _____

Conventional unit maps _____

SURFACE FEATURES

Levee ....

Sand or mud area, dunes, or shifting sand ......

Intricate surface area ....

Gravel beach or glacial moraine ....

Tailings pond ....

VEGETATION

Woods ....

Scrub ....

Orchard ....

Vineyard ....

Mangrove ....

MARINE SHORELINE

Topographic maps:

  Approximate mean high water ....

  Indefinite or unsurveyed ....

Topographic-bathymetric maps:

  Mean high water ....

  Apparent (edge of vegetation) ....

COASTAL FEATURES

Foreshore flat ....

Rock or coral reef ....

Rock bare or awash ....

Group of rocks bare or awash ....

Exposed wreck ....

Depth curve; sounding ....

Breakwater, pier, jetty, or wharf ....

Seawall ....

BATHYMETRIC FEATURES

Area exposed at mean low tide; sounding datum .

Channel ....

Offshore oil or gas: well; platform ....

Sunken rock ....

Figure 9-1. *(Continued)*

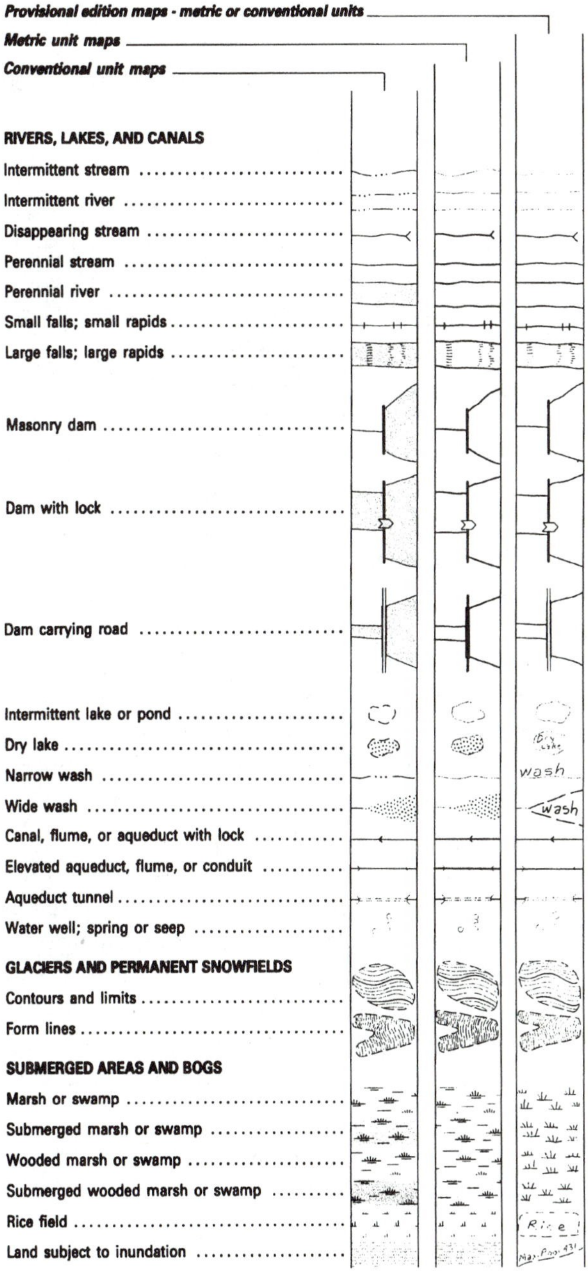

Provisional edition maps - metric or conventional units

Metric unit maps

Conventional unit maps

RIVERS, LAKES, AND CANALS

Intermittent stream ....

Intermittent river ....

Disappearing stream ....

Perennial stream ....

Perennial river ....

Small falls; small rapids ....

Large falls; large rapids ....

Masonry dam ....

Dam with lock ....

Dam carrying road ....

Intermittent lake or pond ....

Dry lake ....

Narrow wash ....

Wide wash ....

Canal, flume, or aqueduct with lock ....

Elevated aqueduct, flume, or conduit ....

Aqueduct tunnel ....

Water well; spring or seep ....

GLACIERS AND PERMANENT SNOWFIELDS

Contours and limits ....

Form lines ....

SUBMERGED AREAS AND BOGS

Marsh or swamp ....

Submerged marsh or swamp ....

Wooded marsh or swamp ....

Submerged wooded marsh or swamp ....

Rice field ....

Land subject to inundation ....

Figure 9-1.   *(Continued)*

will not be to scale. They are for reference only. Be sure to check the legend if there are any symbols that you are not sure of. As with scale, some companies standardize their symbol library, so check with the proper authorities as necessary.

Accuracy

Topo maps are usually not considered to be extremely accurate when compared to other forms of construction graphics. This is because topo maps are most often used to plan earth-moving activities, to determine hydrologic information such as the effect of a certain amount of rainfall over a given area, or other relatively inexact-measurement activities.

For example, when using a topo map to plan earth-moving activities, quantities of earth to be moved are usually estimated by the cubic yard. Since large earth-moving equipment can move many cubic yards of earth in a very short period of time, the cost of earth moving per cubic yard is relatively low. Also, it is almost impossible to account for every small depression or uplifted area on the surface of a job site, so precise measurement of cubic yardage is impossible. However, experience has shown that properly executed topo maps will result in an averaging of the surface highs and lows, therefore offsetting most of the small irregularities.

In most topo surveys, elevations are read to the nearest tenth of a foot. The rod holder or instrument person can take elevation readings at this level of accuracy very quickly, thus saving a great deal of time in the field. Also, readings in tenths of a foot are not as cumbersome to manipulate in the office during the actual map-drafting process. Depending on the size of the area under study and the level of accuracy needed, the readings in tenths of a foot are often rounded to the nearest foot during the drafting process.

The exception to the accuracy level of one tenth of a foot occurs in the case of existing structures. Old foundations, pipes, and so forth may be critical to new construction activities or they may be used as benchmarks. Therefore, the elevation of these items is usually noted to the nearest thousandth of a foot, if possible.

Linear distances in topo data-gathering activities are seldom measured to the degree of accuracy possible with the equipment being used. For example, in measuring down a slope, tape breaking techniques are used. However, the tape may not be tensioned nor temperature corrections calculated as they would be in more-precise measuring work. Readings to the nearest inch are usually more than sufficient. Again, the exception would be linear measurement involving existing structures.

Horizontal angles are usually measured to the nearest degree. Vertical angles for stadia measurement along slopes are usually read to the highest level of accuracy of which the instrument is capable.

Of course, if electronic equipment is used, much more accurate distances and angles can quickly and easily be observed and recorded either manually or by the electronic field book method.

Deciding on Which Surface Features to Show on the Map

Since a single topo map cannot show all the many features of a particular piece of property, some thought must be given as to which features should be observed, measured, and recorded for map production. The decision as to which features are of importance and should be shown on the map depends on the intended use of the map. Example: A topo map of a small residential lot may show all trees in excess of 6 inches in diameter. However, a large industrial plant site which is to be cleared would not show any trees; there would only be a note that the area is wooded. It is often the practice to show only limited features on a large scale map and show more detail in an enlargement of a particular area. Depending on the scale to which the map of the enlarged area is drawn, the smaller area may be shown as an inset on the large map or it may be drawn on a separate sheet.

The following is a list of the most commonly included details on topo maps.

1. Contour lines showing elevations, both existing and proposed. The subject of contour lines is covered in detail below.
2. Major topographic features such as stream beds, hill tops, ridge lines, the edge of cliffs, shore lines, etc.
3. The edge, center line, and intersection of roads, dams, etc. The beginning, center, or ending points of bridges, culverts, and other major structures on roads.
4. The corners of all major building foundations, observation towers, pump stations, etc. The center point of smaller objects such as utility poles, manholes, fire hydrants, etc.
5. Any known socio-geographic features such as city limits, county lines, etc.
6. Utility lines, pipe lines, known easements, etc.
7. Any additional significant feature which will enhance the map maker's ability to draw the map and the reader's ability to read and use the map for its intended purpose.

Illustrating Elevations on Topo Maps

Map drafters have developed several different methods of showing various elevations on topo maps. Several of these methods are shown in Figure 9-2.

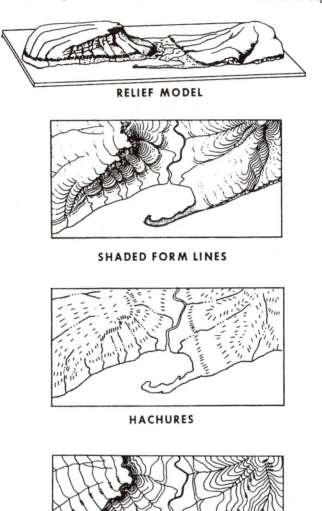

RELIEF MODEL

SHADED FORM LINES

HACHURES

CONTOURS

Figure 9-2.   Methods of showing contours. (Courtesy of U.S. Navy training manual.)

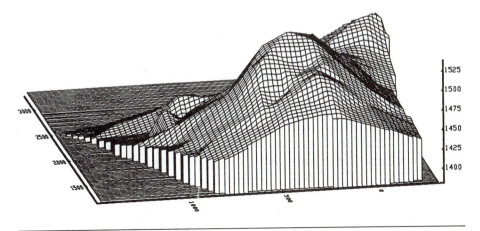

Figure 9-3. A computer-drawn "wire frame" showing topographic elevations. (Courtesy of PAC-SOFT.)

Computer software has been developed which allows map drafters to show terrain features in a precise three-dimensional format. Figure 9-3 shows the result of a computer plot in "wire frame" format.

However, the most widely used method of showing differing elevations on topo maps is the use of contour lines. Whether drawn manually from conventional field notes or by computer from electronically gathered data, the contour lines on a topo map will readily tell the map reader something of the characteristics of the existing terrain. If both existing and proposed contour lines are present, the map reader will be able to mentally picture not only existing terrain conditions, but also how the project site should look upon completion of construction.

Perhaps the easiest way to understand contour lines is to relate them to the surface of a lake located in hilly terrain. Assuming that the surface of the water is perfectly still, the elevation of the surface of the water will be constant everywhere it touches the shore. If a line representing the shoreline were drawn on a map of the area, the result would be a wavy line winding around through the hills and valleys, but ultimately coming back to the point where drawing began. For this discussion, an arbitrary normal elevation of 100 feet will be assigned to this line.

Now imagine a period of dry weather in which the level of the lake fell 5 ft. If a new line were drawn on the same map indicating where the water now touches the shore, this second line would be **inside** the first and would represent the elevation of 95 ft.

If it began to rain and the lake level rose 10 ft, a third line could be drawn representing the shoreline. The third line would be **outside** the first line and would represent an elevation of 105 ft.

An examination of the three lines representing the shoreline of the lake at three different levels (95 ft, 100 ft, and 105 ft) would find that the three lines were not always the same distance apart on the map surface. This might seem strange, because the lines were drawn when the vertical lake levels were exactly 5 ft apart. However, the horizontal distances between lake levels measured on a gently sloping beach and the horizontal distance measured on a steep hill would account for this. On the beach, a 50 ft horizontal measurement might be necessary to account for 5 ft in vertical difference. On a steep hill, a 10 ft horizontal measurement might cover a 5 ft vertical change in elevation. These two different horizontal measurements would account for the shoreline lines being different distances apart on the map.

The rule governing topo lines is: The closer together the lines, the steeper the terrain; the farther apart the lines, the flatter the terrain. Figure 9-4 shows a recommended topo line interval for maps of different scales.

When topo maps are used in construction work, the purpose is usually to show the contractor what the terrain at the project job site looks like in its present condition and what the client wants the job-site terrain to look like

TYPES OF TOPOGRAPHIC MAP	NATURE OF TERRAIN	RECOMMENDED COUNTOUR INTERVAL IN FT.
LARGE SCALE	FLAT	0.5 OR 1
	ROLLING	1 OR 2
	HILLY	2 OR 5
INTERMEDIATE SCALE	FLAT	1, 2, OR 5
	ROLLING	2 OR 5
	HILLY	5 OR 10
SMALL SCALE	FLAT	2, 5, OR 10
	ROLLING	10 OR 20
	HILLY	20 OR 50
	MOUNTAINOUS	50, 100, OR 200

Figure 9-4. Suggested contour intervals for topographic maps. (Courtesy of U.S. Navy training manual.)

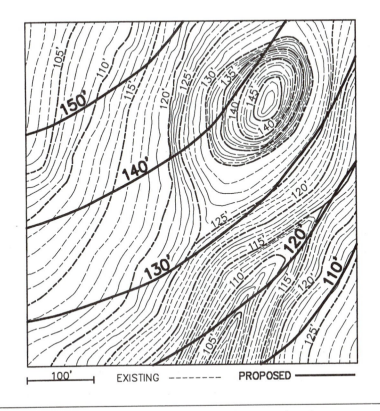

Figure 9-5. Topographic map showing existing and proposed contours.

after the project is finished. This is done by using two types of lines on the topo map: existing contour lines and proposed contour lines. Figure 9-5 shows a job site with the existing contour lines shown as dashed lines and the proposed contour lines shown as solid bold lines.

Note that there are secondary existing contour lines that show the contours in one-foot intervals while the main contour lines show the contour in five-foot intervals. Observe also the method by which low and high areas are shown.

9-3 Collecting Data for Topo Map Production

Several different data-gathering techniques may be employed when preparing to produce a topo map. The techniques chosen will depend on the intended use of the map and the level of accuracy required. It is the responsibility of the field engineer to determine which techniques and procedures are appropriate for each situation.

Preparation for Data Collection

Prior to proceeding with topo map data-collection activities, there are several things the field engineer should know and do. To begin with, a thorough understanding of the purpose of the field activities and the proposed use of the map are paramount. Expedient but safe field procedures, as well as generally accepted and proper field engineering techniques, must be reviewed and practiced. To assist the field engineer in preparing for topo map data-collection activities, the following list of precautions and items to check is presented.

1. Ascertain the purpose for which the map is to be used by consulting with the engineer or other authority who requested that the map be produced. The field engineer is in a much better position to make sound judgments in the field if he or she knows the expressed purpose of this particular map. Example: On a 35-acre plant site, a 10-inch pipe is to be installed underground, connecting one building to another. The buildings are 100 ft apart. In this case, only the area in which excavation is to be done and perhaps several feet on either side of the proposed excavation area need be surveyed. There would certainly be no need to survey the entire 35-acre site. Therefore, much time and money could be saved.

2. Determine whether any prior surveys or engineering studies have recently been made of the area. It may be possible, and perhaps even advisable, to use pre-established benchmarks. Some research and phone calls to various government agencies may be necessary to determine whether government benchmarks have been placed in the area. When no usable existing benchmarks can be found and when extensive construction or further mapping is planned for the site in the relatively near future, it may be advisable to spend extra time in establishing a more permanent benchmarking system than would normally be used for a superficial or partial topo survey.

3. Determine whether the property corners and boundaries are clearly marked or can be easily located. Caution should be used so as not to trespass on adjacent property.

4. Determine whether it is permissible to do any superficial clearing of brush to aid in the survey. Brush clearing may be restricted in some environmentally sensitive areas.

5. Obtain the proper equipment with which to do the job. Check the equipment before leaving for the job site. Be sure to take along the proper safety gear, notebooks, spare batteries, maps, and so forth. Review Chapters 2 through 6 of this text as necessary.

6. Assemble the surveying crew and brief each crew member as to the general aspects of the job. Topo mapping can be done by one person under the ideal conditions and using the proper equipment. However, a crew of three is usually required. This consists of a rod holder, instrument operator, and a recorder or record keeper. More personnel may be added to the crew if extensive linear measurement is required. Of course, more than one crew may be working on the same site in some cases. Their activities must be coordinated by the chief field engineer. If extensive clearing must be done, clearing crews are used to supplement the field engineering crew(s).

7. Upon arrival at the site, confirm that the correct piece of property has been found and the boundaries have been identified. Perform a preliminary observation of the property. Note the general topography of the surface and any safety hazards that may exist. Take precautions as required. Brief the entire crew on the job requirements, safety precautions, communication procedures, specific individual responsibilities, accuracy requirements, and so forth.

8. If previously established benchmarks are to be considered for use, locate them at this time. Look for obvious markers, such as flags and stakes. Follow fence lines and look for corner markers. Look for government markers on exposed parts of permanent structures such as bridge abutments, dam gates, etc. If existing benchmarks are found, note this fact in the field notes. Always use existing benchmarks with caution. It is a good idea to confirm the relative elevations if more than one existing benchmark is to be used. Be alert for damage to existing benchmarks.

9. If no benchmarks exist, establish a relative benchmark for the sole use of the topo survey. Spikes in utility poles, the tops of fire hydrants, marks on concrete curbs, manholes, street intersections, or locations on other permanent structures may be used for this purpose. If no convenient place for a relative benchmark exists, it may be necessary to pour a small concrete monument or to drive a pipe or pin into the ground for use as a benchmark. Place such markers in locations which will be out of the way of future construction work. Assign a relative elevation to the relative benchmark. It is suggested that a relative el-

evation of 100.0 ft or greater be established so that no negative elevations need be recorded for positions below the elevation of the relative benchmark.

10. Determine the best position for control lines. Consider visibility along the length of the line and visibility back to the relative benchmark. The control lines should be as long and as level as possible. Maximum visibility of the terrain on either side of and perpendicular to the control line is desirable. Visibility of major natural features such as ridge lines and stream beds should be optimum. Any large man-made objects should also be visible from a control line. Consider establishing control lines along the property boundaries if necessary. Note the location of the control lines in the field notes.

11. Determine the location of secondary benchmarks if necessary. A secondary benchmark should ideally be visible from the primary benchmark, although this is not absolutely necessary. Tertiary and additional benchmarks may be established as required. If possible, place secondary benchmarks where two or more benchmarks are visible from any one benchmark. This allows for more precise location by the triangulation method and will often prevent gross errors. Precise measurement techniques should be used when establishing benchmarks.

Field Techniques for Data Collection

Methods of collecting field data for topo map production fall into two general categories. These are the **direct contour method** and the **grid method.** Both methods have advantages and disadvantages. Many variations of the two methods have been developed by individual field engineers to meet special needs. In some cases, both methods may be used in different areas on the same job site or in the same areas at different times.

The direct contour method is most often used in preliminary surveys or in surveys of large tracts of land. This method is fast, but it does not provide a great deal of detail.

The grid method is very accurate and provides a great deal of detail for the map maker. However, it is a very slow process compared to the direct contour method.

Field data collection using the direct contour method. Of the two most commonly used methods of field data collection, the direct contour method (sometimes called the "radial" method), is the fastest but provides the least amount of detail for the map maker to use. Generally, this method calls for the field engineer to establish various control points and lines on the property under study. The elevations, angles, and relationships of the control points and selected points on the control lines and other strategic points on the job site are recorded. Figure 9-6 shows the layout of a job site using the direct contour method to gather preliminary data which can be used in planning.

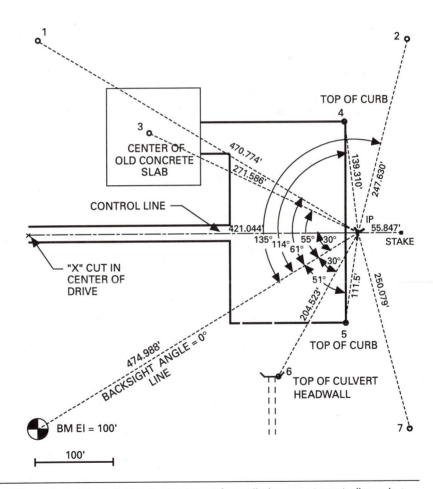

Figure 9-6.   A typical field data-gathering layout for preliminary contour studies using the radial method.

The resulting map is relatively sparse, with only the location and elevations of major features being shown. Depending on the intended use of the map, either elevations or distances may be eliminated. If elevation contours are shown, they must often be interpolated to a high degree by the map maker. This results in a somewhat inaccurate map as far as contours are concerned. Therefore, maps drawn by this method are not often used for planning earthmoving activities or in hydrologic studies.

To begin data collection using the direct contour method, the field engineer must not only have an understanding of the specific requirements of the map makers, but he or she must also be able to determine the most effective and expedient method of gathering the information from a specific piece of property. Of course, each case will be different and experience and practice will aid in perfecting the techniques. However, both experienced and novice field engineers may expect to be sent back to the site of the survey to obtain additional information in some cases. A review of Chapters 2 through 6 is recommended for the novice field engineer.

The following is a list of general procedure rules for use in gathering the field data using the direct contour method. The use of a transit is assumed, but other equipment may be substituted as appropriate. Field notes may be either written manually or entered in an electronic field book. Variations of this technique may be used by experienced personnel where appropriate.

1. Complete the applicable parts of the check list under preparation for data collection above.
2. Set up a transit over a secondary benchmark or over a suitable point on a control line. (Set a hub stake or otherwise mark the instrument location under the instrument if the instrument is not over a benchmark.) Observe a known, established, or relatively primary benchmark, thus establishing a backsight elevation. Record this information in the field notes.
3. Measure and record the distance between the instrument and the primary benchmark. Use taping, pacing, stadia, or EDM, depending on accuracy requirements and the equipment available. Record this information in the field notes.
4. Select the points to be observed from this first position and have the rod holder set up on each of these points in turn. The selection of the observation points is often at the discretion of the field engineer. Commonly used points are foundation corners, ridge lines, creek bottoms,

utility poles, center line of roads, tops of curbs, and so forth. At each point, record the following information in the field notes:

A. The distance from the instrument to the point.

B. The angle between the point and the backsight point.

C. The elevation at the point.

D. Description of the point (i.e., ridge line, SW corner of old foundation, edge of pond, C/L of road, etc.).

5. Proceed to other locations on the primary and other control lines as necessary. If possible, the distance from the primary benchmark to one of the points should be taken to allow for a triangulation check. Sufficient elevations, angles, and distances should be gathered to meet the objectives of the survey, thus allowing the map drafters to produce a topo map which will be of use to the engineers and other planners.

6. Return to the office and either proceed with drafting the map or turn the field notes over to the map drafters or others in authority. Be prepared to answer any questions the map drafters or others might have. Clean and store the surveying equipment.

NOTE

In some cases, the field engineer may direct the rod holder to proceed up or down a slope until an even elevation is found. In other cases, the rod holder is directed to follow the elevation across the terrain while the instrument operator takes an angle and stadia measurement at various points. This method of finding contours is sometimes called the "trace contour" method of data gathering. This is because a particular contour elevation is simply "traced" across the terrain. This method, while somewhat time-consuming in the field, eliminates the problem of elevation interpolation in the office.

Field data collection using the grid method. The grid method of field data collection requires that the field engineer lay out a grid on the surface of the property to be mapped. The size of the squares in the grid is dictated by the general topographic conditions of the area to be mapped and the degree of accuracy needed for the map. While more time-consuming, the grid method relieves the field engineering crew of much of the responsibility of determining which points on the property are critical and must be measured and which are relatively unimportant to the engineers, planners, and map drafters. Figure 9-7 shows a grid layout on a piece of property.

Usually, the grid is laid out with at least one side parallel to a property line and one grid intersection on a property corner or primary benchmark. If the job site is irregular in shape, it is often easier to lay out the grid along magnetic lines or any other straight reference line. Since all the squares in the grid are usually

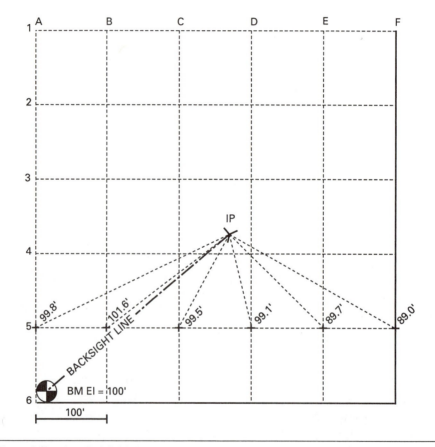

Figure 9-7. A typical field data-gathering layout showing the backsight and several frontsights using the grid method.

the same size, the location of each grid intersection is then quite easily found and recorded. The elevation at each grid intersection is taken and recorded.

The grid method is somewhat slower than the direct contour method because of the time required in laying out the grid itself. The grid layout does not have to be precise and can usually be laid out by taping, using range poles, and maintaining line by eye. Transits may be used to establish the grid lines if a higher level of accuracy is desired.

The grid method has some disadvantages in that it assumes that almost 100 percent of the property is visible and accessible. If the property is heavily wooded or has bodies of water on it, the grid method becomes increasingly difficult, if not impossible. The field engineer would then have to revert to the direct contour method or rely on other methods such as aerial photography and photogrammetrics to provide topo map information.

The following is a list of general procedure rules for use in gathering the field data using the grid method. As previously noted, the use of a transit and a conventional field notebook is assumed. However, other equipment may be substituted when appropriate. Variations of this technique may be used by experienced personnel.

1. Complete the applicable parts of the check list under preparations for data collection above.
2. Select a property line or establish a control line on the property from which the grid layout may be started. Select a position for the transit and record the location in the field notes.
3. Assuming that the required grid size has been determined, measure along the control line or boundary line laying out the grid intersection points. The transit may be used to maintain line if necessary.
4. When the first grid layout line has been marked from one end to the other (continuing all the way to the property lines or until the line of sight is obscured), lateral grid lines may be laid out using one of two methods. One method is to stake a line parallel to but some distance from the first line. The points on each line are used to maintain alignment while staking the lateral lines. The second method is to set the transit over each point on the first line, turn 90 degrees to the line and set stakes on the lateral lines using the transit for alignment.
5. Using the transit setup point selected earlier or the points on the control line, take a backsight on the job-site benchmark. Record this elevation in the field book.

NOTE

It is customary to number the grid lines along one edge and letter them across a lateral edge. This makes field book data recording relatively easy. For example, B6–378.4′ would mean that the elevation at the intersection of grid lines B and 6 would be 378.4 ft relative to whatever benchmark was being used.

6. Have the rod holder proceed to each grid intersection. Take an elevation reading at each intersection. Record each as a frontsight. Each frontsight will have the same backsight reading (see step 5 above) until the instrument is moved. Record the elevation of each intersection in the field book.

NOTE

It is permissible and sometimes advantageous to take elevation readings at the grid intersection points on the lateral lines as they are established. This may save considerable time, but usually requires another person to keep field notes during the process. A sight on the primary benchmark is the backsight and each successive grid intersection is a frontsight.

It is often necessary to establish secondary control lines and benchmarks due to visibility across the property. Extreme caution must be taken when establishing these secondary points and lines. Elevations, angles, and distances must be carefully taken and recorded to establish the relationship of the secondary controls to the primary controls.

Depending on the accuracy requirements, the process of grid layout may be accelerated by simply pacing along a grid line until the approximate location of the intersection is found. The instrument operator may direct the lateral movement of the rod holder to preserve the line and periodically confirm the approximate distances by stadia measurement. Elevations are taken as usual at points found by this method. On large tracts of land where conditions permit, automobiles, all-terrain vehicles, and even horses may be used by the rod holders to cover the area quickly. This is particularly advantageous in areas that are swampy, brushy, or where tall grass is growing. Even small boats have been used by rod holders to cover marshy areas.

Remember, when using the grid method, the elevation and location of all prominent features on the site should be recorded. For example, if a road crosses a line between grid intersections, the location and elevation of the edges and center line of the road should be recorded.

9-4   Producing Topo Maps from Field Data

It is not uncommon for the field engineer to be responsible for drafting the topo map after he or she has collected the field data. Of course, computer graphics programs may be used to draw maps quite easily and should be used if available and if the field engineer is trained in their use. However, there are still many company offices and job sites where computers and/or the proper software are not available, making it necessary to use conventional drafting methods.

Regardless of the method and equipment used, the idea is to get the field data onto paper. Experienced computer graphics users will quickly see how this method can be adapted to computer drafting. Many computer drafting programs have a "layer" system that lends itself to this method quite well.

Below is a step-by-step process used to produce a topo map from field notes.

Step 1. Lay out the grid for the map. This grid must be scaled to the dimensions of the map on which the topo lines are to be drawn. If the map which is to be used is scaled to 1 inch = 50 ft and the grid laid out in the field was 100 ft on a side, the grid should be drawn 2 inches on a side. Draw the grid on a blank piece of paper the same size or slightly larger than the map.

Step 2. Using the field notes, label the grid lines and transfer the grid intersect elevation information to the grid. Transfer any additional information as well, including elevation and location of structures, roads, benchmarks, and so forth.

Step 3. Find the point between each grid intersection where an even elevation line would cross the grid line. Of course, in fairly level terrain, this may not occur between every intersection. In very rough terrain, there may be several even elevation lines crossing between two intersect points. To properly locate the crossing points requires some interpolation as well as some

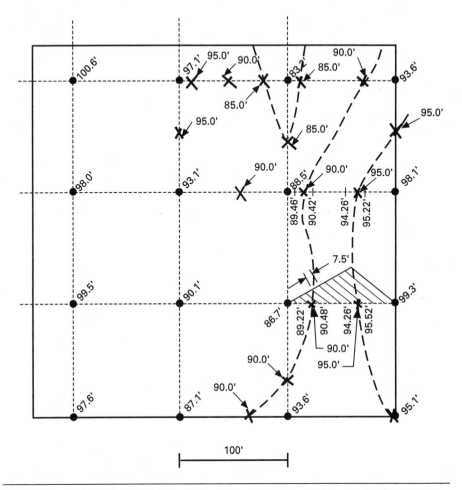

Figure 9-8. A partially completed contour map showing several methods of interpolating contour line positions.

educated guessing. Figure 9-8 shows one method which may be used to estimate the placement of the crossing points. Mark the crossing points with a heavy "X" and label the point with the appropriate elevation.

Step 4. Draw in the elevation lines. This is done by connecting the points where the elevation lines cross the grid lines. A french curve, ships curve, or a similar technique can be used if a computer is not used to do the drawing. Label the elevation lines appropriately.

Step 5. Lay the map over the grid. Align benchmarks, index points, structure lines, control points, and so forth to ensure that the grid is properly located under the map. Using a light table if necessary, trace the topo lines onto the map. Be sure to label the elevations correctly. Note that existing elevation lines are shown as lightly drawn or dashed lines. Note the elevations at points on structures or roads as recorded in the field notes.

Step 6. Check your work. Make several random checks of elevation points on the finished map. If possible, return to the field and review the general shape of the terrain as compared to the map. Compare map elevations with site elevations.

NOTE

People who do a lot of topo map drafting using conventional equipment often have grids of various standard scales drawn on the underside of mylar sheets. The field data and lines are drawn on the top of the mylar using a grease pencil or other erasable medium. The lines and points are then transferred to the map. Later, the lines and points may be erased and the mylar is ready for another use.

Of course, the grid can be drawn directly on the map and that is often done. When this method is being used, draw the grid lines lightly and do the interpolation of the elevation line points on a separate sheet of paper. This keeps the extraneous marking on the map to a minimum.

9-5   Common Uses of Topo Maps in Engineering and Construction

As previously mentioned, topo maps certainly have many uses, but calculating cuts and fills, drawing land profiles, and determining the toe and top of slopes in cut and fill areas on roads and other dirt-moving projects are probably the most common uses in construction work. Of these, the field engineer probably uses the profile drawing technique most often. This technique is often quite useful in visualizing the earth work that must be done in a certain area and in verifying the cuts and fills on roads.

To generate a profile of an area depicted on a topo map is really quite simple and, as with other uses of topo maps, lends itself readily to computer use.

Below is a step-by-step process for generating a profile of a piece of land using a topo map.

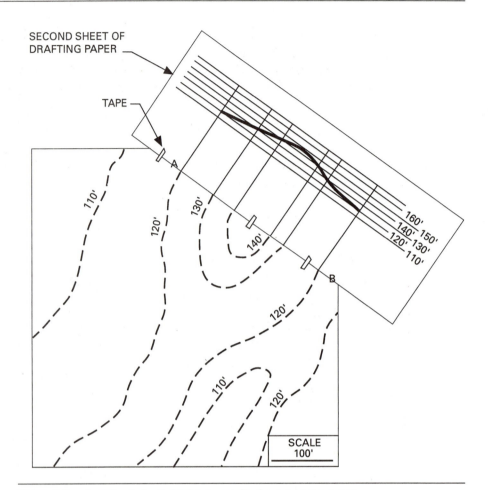

Figure 9-9. A method of establishing a profile from a contour map.

Step 1. On the topo map, locate and draw a line where a profile is to be generated. Figure 9-9 shows a topo map with line A-B drawn for the profile line.

Step 2. Beginning at every point where an elevation line crosses the profile line, draw an extension line onto another sheet of paper temporarily attached to the map. These extension lines must be 90 degrees to the profile line.

Step 3. Draw elevation lines across the extension lines drawn in step 2. These elevation lines must be 90 degrees to the extension lines (parallel to the profile line). Space the elevation lines at the same scale to which the map is drawn. Label the elevation lines according to the elevations shown on the map. This area is known as the profile section. See Figure 9-9.

Step 4. From the elevation line/profile line intersect points on the map, follow each extension line until the same elevation line is reached on the profile section. Mark this intersect point. Continue across the profile, marking each point where an extension line drawn from a plan elevation line crosses a corresponding profile elevation line.

Step 5. Connect the points on the profile elevation lines using a french curve or similar method.

The method of generating a profile may be varied and used in many ways on a construction site. Points on the profile may be projected back to the plan view, and so forth. The profile may show both proposed and existing elevations. The uses are almost limitless. Of course, those trained in the use of computer graphics will quickly see how this method of profile generation can be adapted to the available software.

For an in-depth study of profile and other auxiliary view generation, the field engineer is encouraged to consult texts on engineering graphics and descriptive geometry.

Chapter 9 Study Questions and Exercises

1. What is the best scale for a topo map of a construction job site?
2. Topo maps are usually not drawn to a high level of accuracy, in contrast to other construction graphic documents. Why is this the case?
3. What are the customary levels of accuracy for elevations and angles in topographic survey work?
4. List several features which are normally included on a topo map.
5. What do contour lines on a topo map represent?
6. If a topo map is drawn showing both existing and proposed contour lines, what type of line represents existing elevations and what type of line represents proposed elevations?

7. When gathering field data for a topographic map, should the control line be centrally located on the property or should the control line be located along one of the property boundary lines?

8. Which method of field data collection yields the more accurate data for topo map generation, the grid method or the radial method?

9. What are the advantages of labeling grid lines with numbers and letters?

10. In most instances, even foot elevation points on grid lines will fall between grid intersections. How is the location of the even foot elevation point determined?

10

ROAD, BRIDGE, AND DAM CONSTRUCTION

10-1 Introduction

The construction of roads, bridges, and dams is often referred to as civil construction. The responsibility of the field engineer is not easily defined in civil construction processes due to the diversity of the projects involved. The various sizes and locations of projects, and the number of general contractors and subcontractors involved makes stating anything other than generalities almost impossible.

However, there are enough similarities among projects that a field engineer well-versed in road construction can easily convert to bridge and dam construction, and vice versa, with little difficulty.

The same principles of field layout apply to civil construction as apply to other construction activities. The level of accuracy usually ranges from 1:3000 to 1:5000 for smaller structures and bridges to 1:10,000 for some large, long-span bridges. Of course, accurate records and field book entries are mandatory no matter what the size of the project is.

Due to the nature of civil construction, some unique safety hazards exist. Probably the most hazardous item on a civil construction site is the heavy earth-moving equipment which is in constant use. All workers must be aware of this equipment and the associated hazards.

The field engineer should consult with the company safety representative and the project manager before performing work on any civil site to determine whether any additional site-specific or unusual hazards exist.

10-2   Terminology and Conventions of Road Construction

When roads are planned, the center line is the first reference mark drawn on maps which show where the proposed road is to be built. The proposed center line is established only after many engineering studies have been made to determine the most economic and most appropriate location for the road. The center line for the road may be the center of the paved surface or it may be the center of the median between the lanes of a multi-lane road such as an interstate highway.

After the center line has been established, the road boundaries on either side of the center line are established. These boundaries create what is known as the right-of-way for the road. All construction for the road will take place within the right-of-way boundaries. The distances between the outer right-of-way boundaries varies with the terrain and road design.

The open center area between the lanes of a multi-lane highway is called the median. The area immediately on either side of the paved areas is called a shoulder. The area from the shoulder to the outer boundaries or fence lines is simply referred to as the right-of-way.

Locations along the road must be identified in some manner. This is accomplished by using a system of stations and pluses. A station is an interval of 100 ft and a plus is any distance between 0 ft and 100 ft. Stations can start at any location along the right-of-way, but they usually start at some obvious point. Such a point might be the center of an intersection, a state or county line, or some natural feature like the center of a river channel.

The following is an example of how a location is identified using stations and pluses: A pipeline crosses under a road at station sta. 35+38.92. This means that the pipeline crosses under the road 3500 ft + 38.92 ft or 3538.92 ft from the point where the stationing began. This distance is measured down the center line of the road.

Roads are usually designed to follow the path where the least amount of earth will have to be moved to complete the road. This is simply a matter of economics. However, in most cases, some cutting away of hills and filling in of valleys must be done. These cuts and fills are similar to the grading necessary to prepare a building site for construction. This is discussed in Chapters 6 and 7. Cuts and fills are described by the degree of their slope, the angle of repose of the soil in the cut or fill area, and the location of the top of the cut and the toe or bottom of the fill.

Borrow pits and borrow ditches are often cut in the area of road construction. These are areas where earth is taken from (borrowed) to fill in low areas or to build up the road bed to facilitate drainage. The location and volume of borrow pits is of great importance to the road contractor because the distances that fill material must be hauled will have a significant bearing on the cost of constructing the road.

The same is true of stockpiles of material that have been hauled in or otherwise prepared for use on the road. These stock piles must be planned and monitored to ensure the most economic use of the material and the hauling equipment.

While the material found in the right-of-way can be rearranged to provide a smooth sub-base for the road, the actual paving material is usually placed on base material that has been carefully selected or prepared for this purpose. This material is called select fill because it has been specially selected for filling in the last few inches of the road bed just prior to paving. The select fill will usually have better compaction, water retention, or drainage characteristics than the material that occurs naturally in the area.

The sub-base material just under the select fill material may be compacted and stabilized with chemicals such as lime or cement before the select fill is placed in position.

10-3 Road Layout (Residential)

The layout procedure for roads in residential areas and some sub-divisions is different from road layout in rural areas. In small sub-divisions and more densely occupied areas where some road construction and leveling has already been done, the locations and dimensions of additional roads are frequently referenced to existing structures. In these cases, layout of the road and its curbs, culverts, and so forth, is quite similar to the layout of any structure. A system of linear or radial layout points can be used as described in Chapter 6. In most cases, no major cuts or fills are required.

Stake Location

When standard structure layout methods for roads are used, the offset points are located on 25-ft intervals and at direction change points. The stakes are placed outside the curb or edge of pavement lines. If no curb is to be installed, the offset stakes may be set at the property lines on either side of the street or at some convenient distance from the edge of the proposed pavement, but within the right-of-way for the road. The offset distance will be dictated by the terrain and local custom. If some cutting or filling must be done, the offset stakes must be set outside the cut or fill area so that they will not be disturbed. The field engineer should consult with the project manager to determine the local customary offset distance for pavement edge offset stakes.

Occasionally, the center line of the road is staked for reference and convenience during rough grading. Of course, the road surface is staked for fine grading just prior to paving.

If curbs are to be installed, the standard offset distance is 3 ft from the face of the curb. Because some companies may customarily use a different offset distance, the field engineer should again check with the project manager on this.

Due to the many different designs of curb cross sections, it may be difficult to determine the exact point on the face of the curb from which to measure. In some cases, the flow line of the curb or possibly the back of the curb is used as a reference point for offset stakes as well as other road measurements. See Figure 10-1.

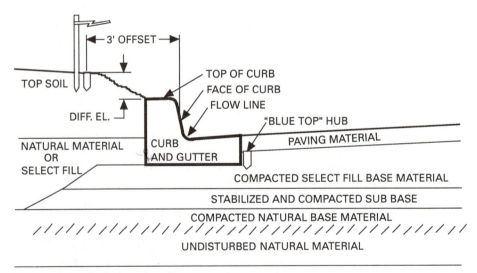

Figure 10-1. Typical road profile.

Stake Elevations

If no curbs are to be installed, the stake elevation is usually set at some distance above or below the finished pavement center line elevation. If curbs are to be installed, the elevation of the stakes is referenced to the top of the curb or the flow line of the curb. The elevation of the hub stake in relation to the road surface or the top or flow line of the curb is, of course, marked on the guard stake.

Other Element Locations

Other elements of the roadway construction, such as cuts in the curbs for driveways, location of culverts and drains, and so forth, are laid out by conventional means. The only difference is in the layout of horizontal curves with a large radius and the layout of vertical curves. The process of laying out these curves is covered later in this chapter.

10-4 Road Layout (Rural)

When roads must be laid out in rural areas where no other construction is in place, the process is somewhat different. The center line of the road is the guiding factor in layout work.

Initial Layout Procedures

Road location may be described by reference to a number of points. Among these are monuments set by the initial surveying party, points on existing structures, or points established by celestial navigation or satellite positioning.

The greater percentage of road layout work today is done with reference to monuments set by advanced engineering and surveying parties or with reference to existing structures or objects. Layout of roads using celestial navigation positioning and positioning by reference to satellites is gaining popularity and may become the standard in the future. However, these methods are beyond the scope of this text and only road layout using established points will be addressed.

Road layout may follow either the radial or the linear methods as described in Chapter 6. Regardless of the method chosen, the initial layout serves to establish rough lines and elevation control points which can be used by the initial clearing, rough grading, and construction crews in their work. The field engineer should keep this in mind during the rough layout work. Many field layout decisions will be based on this fact.

The first consideration is to lay out the center line of the road. This is a simple process if the guiding monuments and control points have been found and identified. The center line is staked at even stations and at any point where a change in direction occurs. Some contractors may wish to have the mid-point or quarter points of stations staked as well. Also, the station points where other structures such as bridges or culverts will be located should be staked. The level of accuracy in laying out the center line of the road in this early stage may be 1:3000 or less.

A word about the stakes and hubs set in this early process. The information regarding the stake position is marked on the stake. This includes the station number, the elevation, cut or fill amounts, and any other information which may be useful at a later date. Elevation hubs may or may not be set at this time. If the terrain is not too drastically different from the finished terrain, hubs may be set. However, if it is obvious that major dirt moving or other engineering must be done in the area, only center line stakes may be set. The remaining hubs and additional stakes will be set at a later date after the clearing and rough grading have been done.

In the early clearing and grading process, hubs may be set only at even stations or at stations closest to bridges, culverts, and other structures. This gives construction crews a benchmark from which to set concrete forms for the bridges, culverts, and other structures. The construction of the bridges and culverts usually begins quite early in the road-construction process because these structures take the longest time to build.

During the placing of the center line stakes, the right-of-way boundaries should be marked as well. This may be accomplished by taping, stadia, or by any other linear measurement technique. It is the custom of some contractors to set the boundary stakes as much as 5 ft inside of the actual boundary line at this stage of the construction process. This tends to prevent inadvertent trespassing onto property outside the right-of-way.

Right-of-way boundary stakes are usually set on lines that are approximately perpendicular to the center line at each station. The boundary stakes are sometimes marked with the center line station number that they correspond to. Of course, boundary stakes set on curves do not have the same station numbers as the center line stakes, so they are usually left blank except for a notation that they are boundary stakes. Ribbons of different colors are used to indicate the center line, the side boundaries, and other structure location stakes.

It is not necessary to use an instrument to establish lines that are perpendicular to center line stations for boundary staking purposes. An estimate of the proper linear location for a boundary stake is sufficient at this point.

One method used to roughly establish a line that is perpendicular to the center line is to stand over the center line facing the side of the road with your arms extended outward from your body at shoulder height. Glance down each arm to ensure that it is aligned with the center line of the road. Next, with eyes closed, bring your arms together in front of your body. Open your eyes and sight between your thumbs to a point near the edge of the right-of-way. Mark this point and measure out from the center line toward it. The boundary stake can be placed on this line after measuring the proper offset distance from the center line. With some practice, you will be able to come to within a few degrees of laying out a perpendicular line using this method. See Figure 10-2.

Figure 10-2.

As the layout of the center line progresses, the center line is extended and laid out all the way to the PI points (if the PI is accessible). (The PI point is the point where the two extended tangent lines of a curve intersect.) See Figure 10-3. When the PI is reached, the included angle between the front tangent and the back tangent is turned, and the layout of the center line continues. The curves are not laid out until later. This prevents the natural cumulative errors which occur in curve layout from following the layout all the way to the end.

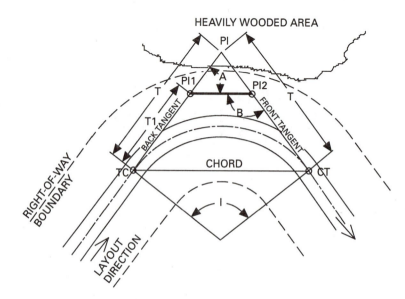

1. Calculate the "T" distance. Then select the point on the back tangent for PI1. Measure along the back tangent line until the PI1 point is reached.
2. Calculate the length of the base of the triangle formed by PI1, PI2, and PI. Use the following formula:

$$\text{BASE LINE LENGTH} = 2 \left(\cos \frac{I}{2} \right) (\text{T-T1})$$

3. Set up a transit over the PI1 point and turn angle A. Angle A may be found by dividing I by 2.
4. Sight along the base of the triangle and measure the distance as calculated above. Set a stake at the end of the base line at point PI2.
5. Move the instrument to point PI2, sight on PI1, and turn angle B. Angle B may be found by: 180-(I/2) CT will be T1 feet from PI2.
6. Continue the traverse on the front tangent.

Figure 10-3.   Establishing a "double PI" on a horizontal curve.

If the PI is not accessible, the field engineer has two choices in laying out the curve area. One choice is to construct an equilateral triangle, with the top at the PI and the base line intersecting the front and back tangents an equal distance from the PI. Figure 10-3 illustrates this procedure. A list of steps to accomplish this procedure is shown as well.

The layout of the back tangent is stopped at the point of intersection of the base line of this triangle, an angle is turned, and a sight is taken along the base of the triangle. The length of the base is measured, another angle is turned to sight forward along the front tangent, and the layout continues. This produces what some field engineers call a "double PI." Extreme care must be taken in recording this procedure in the field notes and on the stakes. It is a good idea to brief the supervisor of the equipment operators as to which curves have double PIs laid out.

The second choice open to the field engineer if the PI is not accessible is to go ahead and stake the curve in the conventional manner.

The placement of boundary stakes usually stops at the TC point and re-sumes at the CT point in the initial layout process. This is because the distance from the tangent line to the boundary line is not constant within the curve area. The placement of the boundary stakes on the curve is done later as the curve is laid out. The layout procedure for curves is covered in detail in Section 10-5.

During the early clearing and grading operations, the field engineer may have to be in almost constant attendance at the job site. Depending on the terrain, factors such as visibility through brush and inaccessible stake locations may cause the field engineer to have to make repeat sightings and re-set stakes after clearing operations and some initial earth moving have been completed.

It should be noted that some landowners in the area of new road construc-tion may be resistive to the construction process. In many cases, the right-of-way may have been taken from the landowners by court action. The owner may be angry and may try to take the anger out on the field engineering crew. In this case, it is better to leave the area immediately and report the incident to your supervisor. It is not the responsibility of the field engineer to deal with the distraught owners of adjacent property.

Intermittent Layout Procedures

After the clearing and any necessary initial grading have been accomplished, a more accurate layout of the road can be started. This phase of the layout can usually be started while the clearing and rough grading operations are pro-gressing on other sections of the road. The secondary layout is done with more accuracy and attention to detail than the first required. The accuracy ratio should meet or exceed the 1:5000 specification. As most modern road layout work is done with EDMIs, this level of accuracy is quite easily obtained.

In the second phase of road layout, the center line may be abandoned as the reference line. This is due to the necessity of moving large quantities of dirt in this area. There may be extensive cutting and filling of the road bed area and stakes would just be in the way. In these cases, stakes and hubs are offset from the center line a convenient distance, perhaps all the way to the right-of-way boundary line.

Another set of stakes may be set in this second phase of layout. These are the cut and fill stakes. These stakes mark the top of cuts and the toe or bottom of fills. The angle of the cut or fill is noted on the stake. The location of these stakes is dependent on the angle of repose of the soil in the immediate area of the cut or fill, the design and construction method associated with the excavation work, and the safety factors that dictate the overall design of the road.

The cut and fill lines are determined by the profile generation method described in Chapter 9 or by mathematically calculating the profile using trigonometry. Neither method is one hundred percent accurate because the consistency of the soil is not always predictable. The cut and fill stakes may have to be adjusted from time to time during construction as job-site conditions change.

In most cases, the cut and fill lines are generated by computer in the office and all the field engineer has to do is locate the station, turn 90° to the center line in the proper direction, measure out from the center line the specified distance to the top of the cut or the toe of the fill, and set a stake.

In the repeated setting of stakes, the elevations on the road bed will be changing. The field engineer must consult with the project manager to determine just which reference elevation he wants used in setting hubs. In one phase of the operation, the supervisor may want hubs set on the edge of the road bed at sub-grade elevation. In another phase, the supervisor may want the hubs set down the center line **and** along the edge at the base elevation. The field engineer must be flexible and read the prints carefully to determine the proper elevations at the various stations.

Final Layout Procedures

Just prior to placing the paving material, the base will be brought to its final elevation. This is the "blue topping" process, as described in Section 6-4. In this process, hubs are set down the center line and on either edge of the road bed and the base material is contoured to its final elevation by referring to these hubs.

After the base material is contoured, seal coats (a sort of primer) and paving material are applied. Some types of paving equipment require that stakes be set with string lines attached in order to guide the paving machine and adjust the depth of the paving material during the paving operation. Others use laser beams for the same purpose. The field engineer may have to assist in setting up the lines or lasers for these pieces of equipment.

After the paving has been completed, the field engineer may have to lay out the road once again in order to place center line stripes, cross walks, signs, and other accessories. Additional shaping of the ditches, median, and culvert entry and exit areas may require additional layout work.

10-5   Laying out Horizontal Curves

Some highway designers say that roads are nothing but a series of curves connected by straight lines. In some parts of the country, this is certainly true. In fact, many roads are more curve than straight line.

It would seem that laying out a curve would be quite simple. One would just go to the radius point and measure outward to the arc line. Repeating this measurement every 25 ft or so along the arc would allow the field engineer to set a series of stakes which describe the curve quite well. This is in fact how a short radius curve is laid out.

However, a curve with a long radius is another matter. In most cases the radius point is simply not accessible. The radius point may be located on private property or it may be geographically inaccessible by being located on a body of water or deep within a mountain.

The inaccessible radius point forces the field engineer to approach curve layout from another perspective. He or she must work along the curve using a transit and tape or EDMIs to accurately lay out the curve.

There are several variations to the methods of laying out horizontal curves, but the method shown here is very accurate and lends itself to field operations quite well. It should be noted that due to rounding errors and other variations in the calculations, the calculated location of points on a curve will vary by a few seconds of arc and a few tenths of a foot from describing a perfect curve. This will usually present no problem to the contractor because the accuracy of the placement of the layout points will be much greater than the accuracy level maintained by the heavy equipment during the construction process.

To understand curve layout one must first know the nomenclature of the various points and lines that are a part of the curve layout process. Figure 10-4 shows a typical curve layout with the pertinent points and lines labeled. The entities shown in parentheses are alternate notations used on some plans and in some texts. Figure 10-4 also gives many of the formulas which are handy to have when dealing with curves.

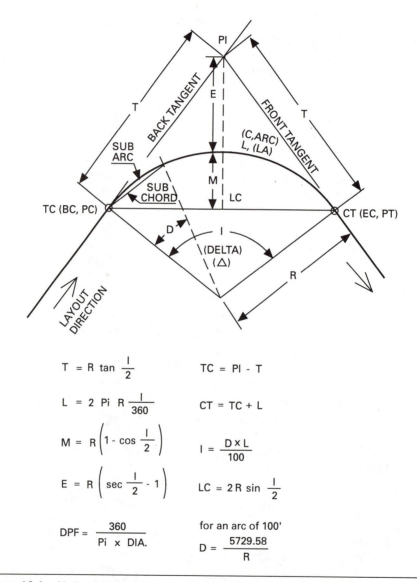

$$T = R \tan \frac{I}{2} \qquad\qquad TC = PI - T$$

$$L = 2 \text{ Pi } R \frac{I}{360} \qquad\qquad CT = TC + L$$

$$M = R\left(1 - \cos \frac{I}{2}\right) \qquad I = \frac{D \times L}{100}$$

$$E = R\left(\sec \frac{I}{2} - 1\right) \qquad LC = 2 R \sin \frac{I}{2}$$

$$DPF = \frac{360}{\text{Pi} \times \text{DIA.}} \qquad\qquad \begin{array}{l} \text{for an arc of 100'} \\[4pt] D = \dfrac{5729.58}{R} \end{array}$$

Figure 10-4.   Horizontal curve formulas and notation.

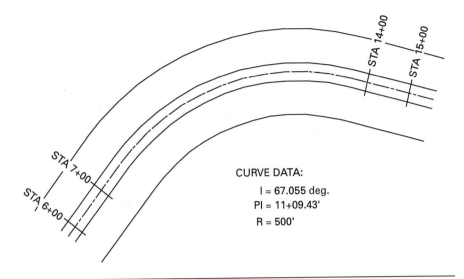

CURVE DATA:

I = 67.055 deg.

PI = 11+09.43'

R = 500'

Figure 10-5. A typical horizontal curve as it might appear on a highway print.

The information given on a set of plans concerning a particular curve will vary because of the preferences of the agencies that prepared the plans. However, the basic information that it will be necessary to have will be the station number where the curve starts (tangent to curve or TC) or the station number of the point of intersection of the tangents (PI), the included angle (I or Delta), and the radius (R). With this information available, the rest of the curve data can be calculated.

Figure 10-5 shows a typical highway curve with only the center line drawn and minimum essential information listed. All extraneous information that would normally be included on the prints for a road has been excluded for clarity.

The following is a step-by-step procedure in which the necessary data is calculated and the curve shown in Figure 10-5 is laid out. It is strongly suggested that the field engineer find a quiet place with a desk or table where he or she can study the plans and do the calculations for the curve layout. Many experienced field engineers draw the curve to scale to verify their figures before going to the field to do the layout work.

Step 1. In the field note book, list the data needed to lay out the curve. This will include I, R, station number at the P I, L, T, the station number

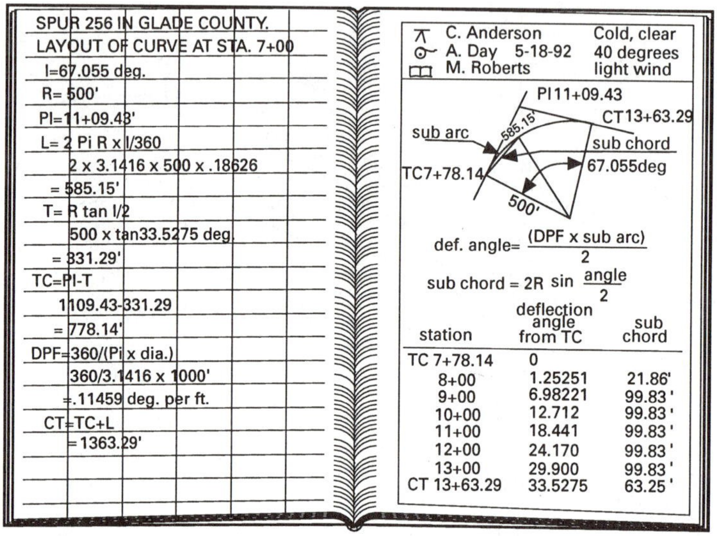

Figure 10-6.   Field book with horizontal curve data.

at TC, the degrees of deflection per foot of arc (DPF), and the CT station number. Refer to Figure 10-4 for the formulas necessary to determine any of the missing data. Perform the necessary calculations and enter the data in the field book.

Step 2. Sketch the curve in the field book. Note the points, lines, and other data on the sketch. Figure 10-6 shows how the field book entries might appear for the curve shown in Figure 10-5.

Step 3. Consult with the project manager about how far apart to place the stakes on the curve. In this example, stakes will be placed at the beginning of the curve (TC), the end of the curve (CT), and on every even station (every 100 ft) on the curve.

NOTE

In rough grading operations, placing stakes every 100 ft may be acceptable. However, stakes may be required every 25 ft or less for finish work. A shortcut method for placing intermittent stakes is explained later in this section.

Step 4. List the stations where stakes are to be set in the field book. Calculate the deflection angle from TC to the first stake. Calculate the length of the sub chord from TC to the first stake. The formula and an example of deflection angle and sub chord calculations are shown in Figure 10-7. Note that on curves with a long radius, the sub chord distance and the sub arc distances are essentially the same. Enter this data in the field book.

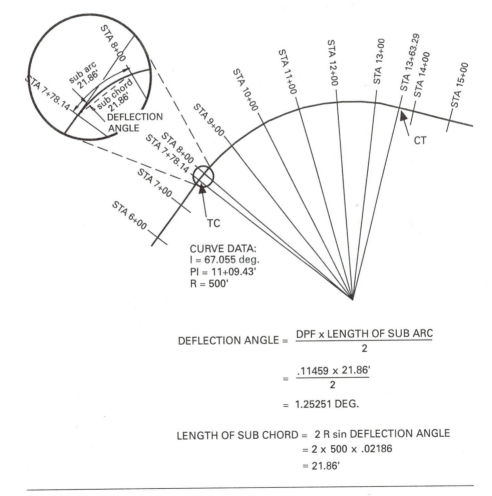

CURVE DATA:
I = 67.055 deg.
PI = 11+09.43'
R = 500'

$$\text{DEFLECTION ANGLE} = \frac{\text{DPF x LENGTH OF SUB ARC}}{2}$$

$$= \frac{.11459 \times 21.86'}{2}$$

$$= 1.25251 \text{ DEG.}$$

$$\text{LENGTH OF SUB CHORD} = 2\,R \sin \text{DEFLECTION ANGLE}$$
$$= 2 \times 500 \times .02186$$
$$= 21.86'$$

Figure 10-7.

Step 5. Next, calculate the deflection angles to the rest of the stakes on the curve that are at even stations. This procedure is shown in Figure 10-8. Enter this information in the field book.

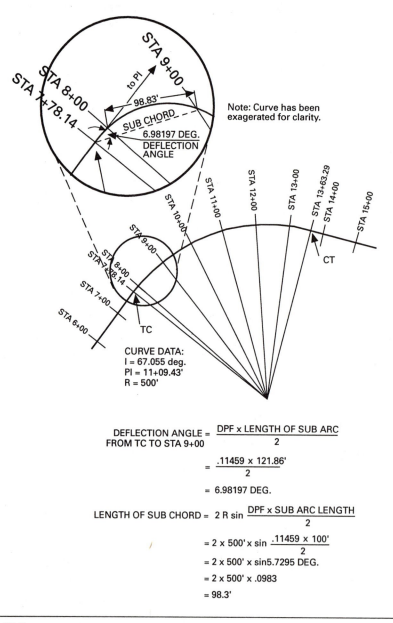

Note: Curve has been exaggerated for clarity.

CURVE DATA:
I = 67.055 deg.
PI = 11+09.43'
R = 500'

$$\text{DEFLECTION ANGLE} = \frac{\text{DPF} \times \text{LENGTH OF SUB ARC}}{2}$$

$$= \frac{.11459 \times 121.86'}{2}$$

$$= 6.98197 \text{ DEG.}$$

$$\text{LENGTH OF SUB CHORD} = 2 \, R \sin \frac{\text{DPF} \times \text{SUB ARC LENGTH}}{2}$$

$$= 2 \times 500' \times \sin \frac{.11459 \times 100'}{2}$$

$$= 2 \times 500' \times \sin 5.7295 \text{ DEG.}$$

$$= 2 \times 500' \times .0983$$

$$= 98.3'$$

Figure 10-8.

Step 6. When the last even station on the curve is reached, calculate the deflection angle and the sub chord for the final part of the curve. The sub arc distance will be CT minus the last even station on the curve. Figure 10-9 shows this last calculation. Make the proper entries in the field book.

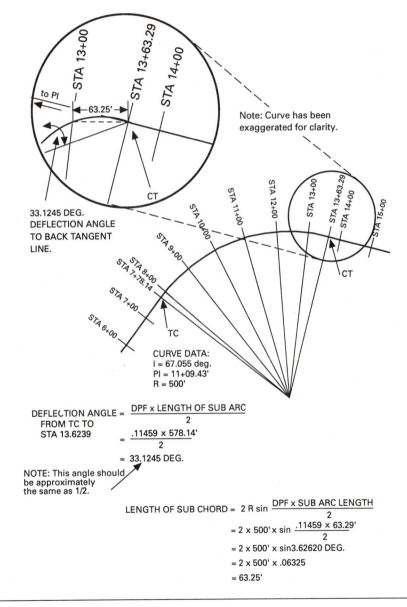

Note: Curve has been exaggerated for clarity.

33.1245 DEG. DEFLECTION ANGLE TO BACK TANGENT LINE.

CURVE DATA:
I = 67.055 deg.
PI = 11+09.43'
R = 500'

$$\text{DEFLECTION ANGLE} = \frac{\text{DPF} \times \text{LENGTH OF SUB ARC}}{2}$$
FROM TC TO STA 13.6239
$$= \frac{.11459 \times 578.14'}{2}$$
$$= 33.1245 \text{ DEG.}$$

NOTE: This angle should be approximately the same as 1/2.

$$\text{LENGTH OF SUB CHORD} = 2\,R \sin \frac{\text{DPF} \times \text{SUB ARC LENGTH}}{2}$$
$$= 2 \times 500' \times \sin \frac{.11459 \times 63.29'}{2}$$
$$= 2 \times 500' \times \sin 3.62620 \text{ DEG.}$$
$$= 2 \times 500' \times .06325$$
$$= 63.25'$$

Figure 10-9.

Step 7. After the calculations have been completed and checked, proceed to the field. To set the first stake on the curve, set up a transit over the TC point. Align the transit scope with the center line of the road, sight on the PI, and turn the first deflection angle to the right or left as required. Depress the instrument scope and stretch a tape from the TC stake along the line of sight of the instrument. Measure off the length of the first sub chord and set a stake.

Step 8. Sight the PI point again. Turn the second deflection angle and measure off the sub-chord distance from the first stake to the second stake. Continue until the curve is staked and the last deflection angle has been turned. **Note:** Always sight the PI before turning each deflection angle. Do not simply add the difference to the previous angle. This will help in avoiding cumulative errors.

Step 9. Review your work. Do the stakes appear to form a symmetrical curve? Is the total of the deflection angles equal to the final deflection angle? Did the last angle and distance measured seem to work out right? Is the last stake on the center line previously laid out? Review the mathematics. Complete the field notes, secure the instruments, and proceed to the next job or report to your supervisor for another assignment.

Setting Intermediate Stakes

One of the easiest ways to set intermediate stakes is to measure laterally off the tape while laying out the sub chords. This takes some calculation, but the results are quite accurate and the process is much quicker than trying to turn a deflection angle for each intermediate stake.

To understand this process, refer to Figure 10-10 while following these steps. This example uses the same curve data as in the previous examples.

1. Calculate the M, M', and the X values as shown in the examples in Figure 10-10.
2. Divide the length of the sub chord into fourths. Stretch a tape along the sub-chord line between the even stations. At the end of the first fourth, measure outward 90° from the tape (toward the curve line) the X distance and stake the location.
3. At the end of the second fourth, measure outward the M distance and stake the location.

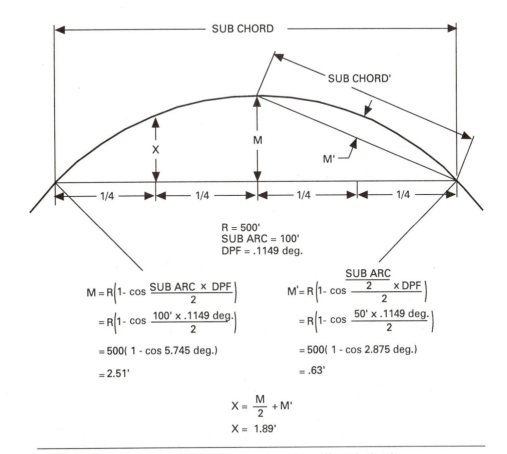

R = 500'
SUB ARC = 100'
DPF = .1149 deg.

$$M = R\left(1- \cos \frac{\text{SUB ARC} \times \text{DPF}}{2}\right)$$

$$= R\left(1- \cos \frac{100' \times .1149 \text{ deg.}}{2}\right)$$

$$= 500(1 - \cos 5.745 \text{ deg.})$$

$$= 2.51'$$

$$M' = R\left(1- \cos \frac{\frac{\text{SUB ARC}}{2} \times \text{DPF}}{2}\right)$$

$$= R\left(1- \cos \frac{50' \times .1149 \text{ deg.}}{2}\right)$$

$$= 500(1 - \cos 2.875 \text{ deg.})$$

$$= .63'$$

$$X = \frac{M}{2} + M'$$

$$X = 1.89'$$

Figure 10-10.   Method of setting intermediate stakes off a sub chord.

4. At the three-fourths point on the tape, measure outward the X distance again and stake the location.

While this method is not precise, it is accurate enough for most curve layout work on highways and other roads.

Moving up on a Curve

In many cases, especially on long curves, many of the stake points toward the end of the curve will not be visible from the TC point. It will be necessary to "move up" on the curve to complete the layout. This is a simple procedure, but it does take some planning.

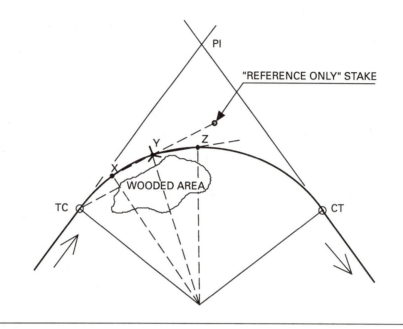

Figure 10-11.

Refer to Figure 10-11. In this figure, the wooded area has made sighting point Z from the TC impossible. To solve the problem, follow these steps:

1. After setting stakes at X and Y, move the transit to point Y.
2. Set up over Y, level the instrument, and sight on X.
3. Plunge the scope or turn 180°.
4. Calculate the deflection angle, sub arc, and sub chord distance from point Y to point Z. Use the formulas in Figure 10-7.
5. Turn the deflection angle to Z, just as you did the first deflection angle, but using the newly calculated data. Lay out the sub chord from Y to Z and set the Z stake.
6. Continue around the curve adding each deflection angle to the previous one and turning each new deflection angle from a sight on the X-Y line.

It may help you to understand this process if you think of the Y point as a new TC point which just happened to fall on an even station. It may also be helpful to set a "reference only" stake on the extended X-Y line to serve as a temporary PI. This eliminates the necessity of plunging the scope or turning 180° each time a new deflection angle is turned.

10-6   Layout out Vertical Curves

The layout process for vertical curves is quite different from the horizontal curve layout process. In laying out the vertical curve, the field engineer is concerned with the distance the curve line is offset from the tangent line at specified station points from the beginning of the curve through the low point or crest of the curve and on to the end of the curve.

Figure 10-12 shows a typical vertical curve with the pertinent points and lines labeled. This figure also includes many of the formulas which are helpful when dealing with vertical curves.

As in the case of horizontal curves, the data contained on a set of plans concerning the horizontal curves will vary. However, the minimum informa-

BVC El. = PVI El. (+ or - l1 x g1)
EVC El. = PVI El. (+ or - l2 x g2)

$$e = \frac{l1\ l2}{2(l1 + l2)}\ \frac{(g^2 - g^1)}{100}$$

$$\text{Tangent line elevation} = \text{tangent line elevation at previous station} + \text{or} - \frac{l1 \text{ or } l2 \text{ to previous station}}{100}$$

Distance from back tangent line to curve line $= (X/l1)^2$ x e where x = distance from BVC

Distance from front tangent line to curve line $= (Y/l2)^2$ x e where y = distance from EVC

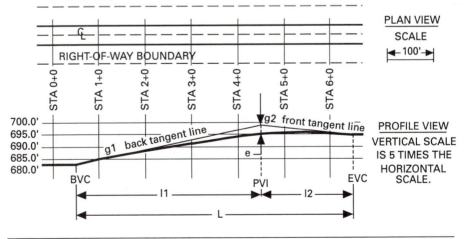

Figure 10-12.   Vertical curve formulas and notation.

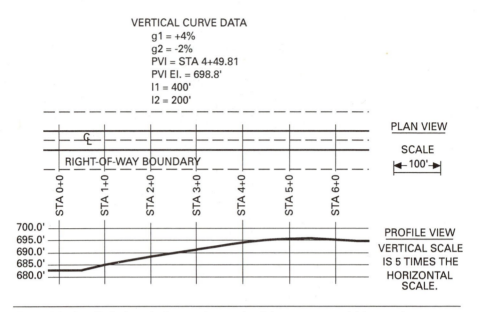

Figure 10-13. Typical vertical curve data as it might appear on a highway print.

tion needed to perform the calculations for laying out the vertical curve are g1, g2, PVI station, PVI elevation, and l1 and l2.

Figure 10-13 shows a vertical curve with all but the essential information removed. This is similar to how the curve would appear on a highway print. It looks rather bare, but after the necessary calculations, all the entities will be identified.

The following is a step-by-step procedure in which the necessary data is calculated and the curve shown in Figure 10-13 is laid out. As in the case of the horizontal curve, it is suggested that the field engineer find a quiet place with a desk or table where he or she can work while studying the plans and performing the calculations prior to going to the field. Also, as with horizontal curves, many experienced field engineers will draw the curve to scale to check their calculations before going to the field to do the actual layout.

Step 1. In the field book, list the data needed to lay out the curve. This will include g1, g2, PVI station, PVI elevation, l1, l2, BVC elevation, BVC station, EVC station, EVC elevation, e, and the necessary formulas to calculate the tangent line elevations, tangent offset distances, and curve point elevations. Refer to Figure 10-12 for the formulas necessary to determine any of the missing data.

Step 2. Sketch the curve in the field book. Note the points, lines, and other data on the sketch. Figure 10-14 shows how the field book entries might appear for the curve shown in Figure 10-13.

Step 3. Consult with the project manager about how far apart to place the stakes on the curve. In this example, stakes will be placed at the beginning of the vertical curve (BVC), the end of the vertical curve (EVC), and on every even station (every 100 ft) on the curve.

Step 4. In the field book, set up a table similar to the one shown in Figure 10-14. In the "Stations for elevations" column list the stations where stakes are to be set.

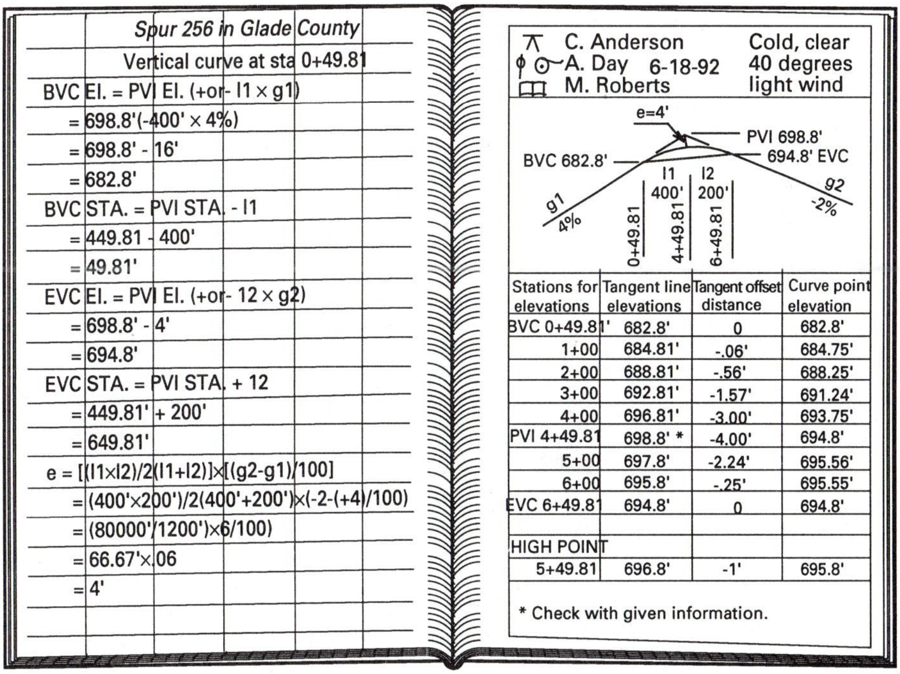

Figure 10-14. Field book with vertical curve data.

Step 5. Calculate the tangent line elevation for each station listed. Refer to Figure 10-12 for the necessary formulas. Enter these elevations in the field book.

Step 6. Calculate the tangent offset distances for each station listed on both tangent lines. Refer to Figure 10-12 for the necessary formula. Enter these distances in the field book.

Step 7. Calculate the curve point elevations for each station listed by subtracting or adding the offset distances to or from the tangent line elevations.

Step 8. Check the data. After the calculations have been completed and checked, many field engineers plot their data to scale to determine whether the curve appears to conform to the data. Figure 10-15 shows a plot of the curve data as shown in the field book in Figure 10-14. In the drawing in Figure 10-15, the vertical scale is five times the horizontal scale. Using different scales is common in road work. It allows the comparatively small changes in vertical distances to be visualized more easily.

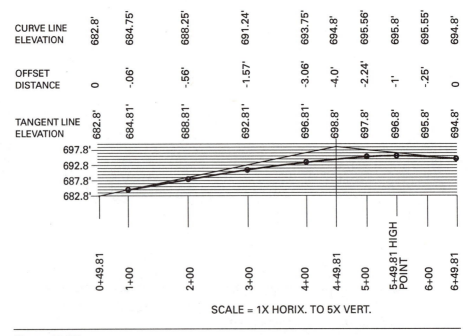

Figure 10-15.

Step 9. Lay out the curve in the field. After checking the data, proceed to the field. Set up a transit on the BVC point and align the scope of the instrument with the center line of the road. This is done by either sighting on the PVI, EVC, or other stake either ahead of or behind the BVC point. With the scope aligned with the center line, proceed to stake and place hubs at the stations as noted in the field book. Use proper taping techniques such as breaking tape, or some other method. See Chapter 4, Section 7, as required.

Step 10. Determine the amount of cut or fill on the curve.

Method 1. After the hubs are in place at each station, the stakes must be marked for elevation. This is done by simply performing a differential leveling operation. With the instrument set up at a convenient location where the most stations are visible, sight on a rod set on a benchmark. Record this as a backsight. Have the rod holder move to each station on the curve and record each station elevation as a front sight. Calculate the existing elevation at each station and record this information in the field book. Mark the guard stake at each hub with the amount of cut or fill required to bring the existing elevation to the curve line elevation.

Method 2. Set up on the BVC point and measure the distance from the BVC hub to the horizontal center line of the instrument. **Note:** This will not be exact, but it is close enough for this work. Align the scope with the center line of the road and elevate the scope to a setting that corresponds to the percent of grade of the tangent line on the curve. Sight on a rod set on the hubs at each station and record this elevation. The cut or fill at each station is determined by:

Cut or Fill = Tangent line offset − [(BVC El.) − (BVC El. + HI − RR)]

where RR = rod reading.

Note: Be sure to change signs as necessary if the curve is dipping or cresting or if the existing elevation is above or below the proposed curve line. Mark the guard stakes with the cut or fill data.

Step 11. Review your work. Is the last stake on the center line as previously laid out? Did the tangent offsets for the last stake fit the calculations for the EVC elevation? Review the mathematics. Complete the field notes, secure the instruments, and proceed to the next job or report to your supervisor for another assignment.

10-7   Determining High or Low Points on Vertical Curves

The field engineer may be required to find the high or low points on vertical curves. This is usually required on dipping curves so that the drainage boxes, other drainage fittings, or curb cuts may be placed in the proper location.

METHOD FOR CALCULATING THE HIGH POINT
OR LOW POINT ON VERTICAL CURVES.

BETWEEN
BVC AND PVI
$$\text{HIGH POINT} = \frac{\left(\frac{l1}{100}\right)^2 \times g1}{2e}$$

BETWEEN
EVC AND PVI
$$\text{HIGH POINT} = \frac{\left(\frac{l2}{100}\right)^2 \times g2}{2e}$$

Procedure: Estimate the location of the low or high point. Select the appropriate formula and try it. If the results are obviously incorrect, try the other formula.
Example: Refer to the curve in Figure 10-13. If you estimate the high point is between BVC and PVI, the calculations would be:

$$\text{HIGH POINT} = \frac{\left(\frac{400}{100}\right)^2 \times 4}{8}$$

$$= \frac{(4)^2 \times 4}{8}$$

$$= \frac{64}{8}$$

$$= \quad 8 \text{ or } 8 \text{ stations from BVC}$$

This is obviously incorrect as the curve does not cover 8 stations. Using the second formula finds:

$$\text{HIGH POINT} = \frac{\left(\frac{200}{100}\right)^2 \times 2}{8}$$

$$= \frac{(4)^2 \times 2}{8}$$

$$= \frac{8}{8}$$

$$= \quad 1 \text{ or } 1 \text{ station from EVC at STA } 5+49.81$$

Figure 10-16.   Formulas and procedures for calculating the high and/or low points on vertical curves.

To calculate the high or low point on unsymmetrical vertical curves, a trial-and-error method is often used. It is not always obvious on which side of the PVI point the high or low point lies. This is particularly true of curves with grades (g1 and g2) and lengths (l1 and l2) which are almost identical. In these cases, two formulas are used. One formula is tried and if that formula does not produce the required results, the other formula is used.

Figure 10-16 shows the formulas and an example of the calculations using the curve in Figure 10-15.

Once the high or low point station has been found, the tangent offset distance and curve elevation at that station may be found using the formulas in Figure 10-12.

It is quite common to combine the properties of horizontal and vertical curves in one curve. While this sounds complicated, it can easily be handled by the field engineer with some study and planning. Simply think of each curve separately and mentally overlay them.

The layout of horizontal and vertical curves is one of the more difficult services the field engineer is expected to perform. Only study, practice, and experience will bring proficiency in proper and timely execution of curve layout. With a bit of thought, planning, and some experience, this will present no appreciable problem to the field engineer.

10-8   Bridge and Dam Construction

Bridges

The layout procedure for bridges is not appreciably different from laying out any large structure, with the exception of the location of control lines and the accessibility factor. The main control line used in bridges is the roadway center line. All other lateral dimensions are taken from this line. In some cases, it may be advisable to establish additional control lines that are offset from but parallel to the center line, especially if the center of the bridge tends to become cluttered with construction equipment and material.

In many cases, control lines may have to be established that are completely off to one side of the bridge to avoid obstructions. Lasers can be set up on these lines and targets used to obtain reference distances from the laser line back to the surface of the bridge.

Longitudinal measurements on bridges are specified by referring to the station number of the road as it proceeds across the bridge. Elevations are taken from benchmarks, just as in other forms of construction.

Radial staking methods may be used, as may lasers, especially if the bridge is to be constructed over a large body of water. Foundation pile-driving equipment may have to be barge mounted, which presents a positioning problem. In these cases, radial positioning using EDMIs or twin lasers for triangulation may be necessary to monitor the position of the pile-driving equipment.

Curved bridges can be laid out in the same manner as other road curves, including horizontal, vertical, and combined curves. However, in most bridge construction projects, the elevations at the various stations along the center line will be dictated by the elevation of the superstructure components. This makes the calculation of the vertical curve elevations unnecessary.

Culverts

Culverts may be constructed of cast-in-place concrete, pre-fabricated concrete components, or metal pipe material. The elevation of square or rectangular culverts is usually specified by referencing the flow line. This is the top surface of the floor of the culvert. The elevation of round culverts is specified by referencing the invert of the pipe. This is the lowest point on the inside diameter of the pipe.

Culverts often run at an angle to the roadway and may or may not be level. Culverts which are set at an angle other than 90° to the center line of the road are said to be skewed. The two terms associated with this angular placement are skew number and skew angle. The skew number is the clockwise-measured angle between the center line of the road and the center line of the culvert. This angle is measured on the side of the culvert with the lower station number. The skew angle is the angle between the culvert center line and a line drawn 90° to the road center line. See Figure 10-17.

Culverts are laid out similar to the way bridges are laid out. The center line of the culvert, the skew number and skew angle, and the elevation of the flow line or the invert are the governing factors. Layout stakes are set on the center lines and offsets as necessary to facilitate construction. Elevation reference benchmarks are often set nearby in convenient locations to facilitate the setting of forms, the grading of the bedding material, and the setting of pre-fabricated components.

The entrances and exits to skewed culverts are often set parallel to the edge of the road. To lay out the entrance and exit lines requires some planning on the part of the field engineer, but should present no particularly difficult problems. The entry and exit exterior form lines are described by setting offset stakes. Batter boards similar to those used in other types of foundation construction may or may not be used, depending on local custom, the forming method, and other factors.

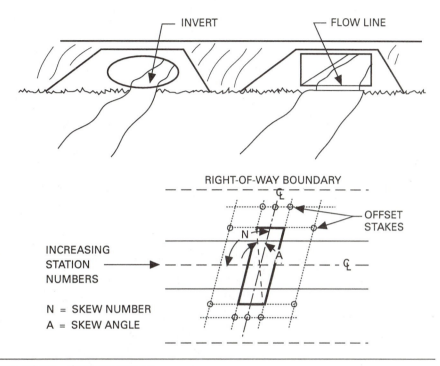

Figure 10-17. Culvert notations.

Dams

The construction of dams is much like road construction. Most dams are laid out on a center line with station numbers. The main elevation reference point is the top of the dam. Almost all relevant elevations on the project will be referenced to the top of dam elevation. This means that most of the elevations on the project will be set below, rather than above, a certain point. The gates, overflows, and other concrete and steel structures are designed and dimensioned similar to any other structure. Both radial and linear staking methods may be used.

The construction of earth-filled dams presents some unique problems. The dam is constructed similar to the fill on a highway, but with different compaction specifications and material requirements. The main difference will be in the type of fill material used. The core of the dam may have one type of fill material while the outer portion of the fill may be of another type. Constructing this vertical "sandwich" of different materials requires that the field engi-

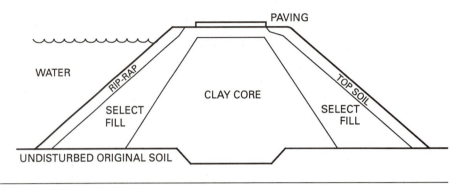

Figure 10-18. Lateral profile of a typical dam.

neer monitor the widths as well as the elevation of the various types of material as construction of the dam proceeds in layers. See Figure 10-18.

The exterior of earth-filled dams is sloped, compacted, and graded in much the same manner as highway fills. The water side of earth dams is usually covered with rip-rap. This is nothing more than large stones which prevent wave action from eroding the sloping surface of the dam. The downstream side is usually covered with a type of soil that promotes good grass growth to control erosion. The rip-rap layer on the water side and the top soil layer on the downstream side must conform to the specifications as to thickness and compaction. It is the responsibility of the field engineer to set elevation stakes and monitor the elevations of the material as it is put into place.

Many dams are curved. The curvature of dams is treated just like the curve of a road. Most dam prints will give the angles, radius, and other information relative to calculating the curve dimensions in much the same way as a highway print.

Chapter 10 Review Questions and Exercises

1. What is the primary reference line associated with road layout?

2. What does the term "right-of-way" refer to in road construction?

3. What is a "station"? What is a "plus"?

4. What is the standard recommended offset distance for stakes when using a curb face as a reference point?

5. Are hub stakes set in the initial survey of a new road?

6. Why do some contractors set boundary stakes some distance inside the true boundary position in the early stages of road construction?

7. After initial rough clearing and grading of the area between the boundaries of a roadway, the center line of the road is not used as the main reference line. Why is this the case?

8. Is it necessary to stake the toe and top of cuts and fills? Why?

9. Why are "blue top" hubs set to the elevation of the top of the base material?

10. Define these curve-related terms: TC, PI, CT.

11. Given R = 250', TC = 3 + 45.87', and I = 35.5°, for a horizontal curve, calculate the values for T, DPF, and the CT.

12. Prepare a sketch of the curve described in question 11 above. Label the various values as necessary.

13. When staking a long curve, stations often are obscured by brush or other items that are outside of the right-of-way. How can the field engineer continue to lay out the curve if the stations are not visible from the TC point?

14. Given g1 = 3.5%, g2 = –2.5%, PVI = 7 + 54.34', PVI El. = 253.86', l1 = 300', and l2 = 400' for a vertical curve, calculate the "e" value, the tangent offset distances, and the curve elevations every 100 ft beginning at the BVC station and continuing through the EVC station.

15. Prepare a sketch of the curve described in question 14 above. Label the various values as necessary.

16. Calculate the high or low point on the curve described in question 14 above.

17. When laying out a culvert, what do the terms "skew angle" and "skew number" refer to?

11

UTILITY CONSTRUCTION

11-1 Introduction

The subject of utility construction is quite broad and covers both above ground and under ground work. This work is carried out by private contractors, municipalities and other government agencies, and the utility companies themselves. The extent of the projects may run from a simple water meter installation to constructing a cross-country power line.

Due to the extreme diversity represented in this division of the construction industry, the field engineer must be more flexible than ever in approaching utility project layout work. The basics will remain the same, but specific requirements of various utility companies and contractors may vary greatly.

This chapter will discuss some of the general conventions associated with both above ground and under ground work. The basics will be stressed, as will the safety aspects of utility construction.

11-2 Above Ground Utilities

Of course, the most common of above ground utilities is the overhead power or communication line. As the construction principles for communication and power lines are almost identical, this section will refer to the power line only.

In the structural design of power lines, engineers must consider the weight and strength of the wire, the height and strength of the poles, the geography of the area, and the general layout of the line route. The climate of the region must be considered as well. Ice and wind loads will dictate much of the design having to do with wire strength and pole spacing.

When the field engineer goes to the field to lay out a power line, maps, specifications, and prints will be provided. It is the responsibility of the field engineer to identify the job site, mark the position for the poles and accessory items, and ensure that the right-of-way is clearly marked.

It may seem ridiculous to point out that there should be any concern over identifying the proper job site, but this often presents a problem, especially in remote areas. This subject is addressed in Chapter 6. A review of these procedures is suggested if there is a question as to job-site identification.

Once the proper job site has been identified, the procedure for laying out an overhead power line is much like that for laying out a road or highway, except that there will usually be no cutting or filling to be done.

Power lines are laid out on a line which is usually in the center of an easement or right-of-way. The right-of-way must be identified and marked to prevent inadvertent trespassing onto adjacent property by utility construction crews. Often, clearing must be done to allow the construction crews to enter the area to erect poles and string the wire.

The field engineer must often be constantly available to provide right-of-way reference lines for the clearing crews. After any necessary clearing has been done, the field engineer must stake the location for the poles and accessories.

Pole Location

In power line construction, one often hears the term "high-tension" lines. This means that the lines are actually stretched between the tops of the poles and a certain amount of tension is placed on the wire. This is done to maintain the wire at a certain height above the ground and to reduce whip or sway in windy conditions.

Because the wires are tensioned, poles must be kept in as straight a line as possible to prevent any side stress from being placed on the pole. Side stress on a pole will eventually cause it to lean or bend. Side-stressed poles are prone to breakage in adverse weather conditions. It is the responsibility of the field engineer to maintain the alignment of the poles within the acceptable standards for each line.

When the location for a pole has been determined, the field engineer will set a stake at the center of the pole location. The drilling-machine operator will line up the center of the bit on the stake and drill the hole for the pole.

Only in rare circumstances will offset stakes be used. The stake for the pole should be marked with information about the pole. The pole number, size, and length should be noted on the stake. Of course, the installation crews will have a set of prints with the same information, but this will serve as a double check. Also, poles and pole hardware may be delivered and distributed along the right-of-way several days prior to being erected. The delivery crews may or may not have a complete set of prints. In either case, the delivery crew can read the pole description on the stake and unload the proper pole and pole hardware at each stake location. No elevation hubs are set.

Laying out a new line which intersects an existing line or installing a new pole in an existing line for any reason requires extreme care in staking the pole location. This is due to the fact that the pole must be placed in alignment with the existing wires. This may be particularly difficult if the visibility between adjacent poles is limited by brush, trees, and so forth. In these cases, it is not uncommon for the field engineer to use a transit to locate the point precisely under the overhead wires.

This may be done in several ways. One way is to set up a transit under the wires and work it to a point directly under the wires, just as you would do if you were working the transit to a line between stakes. (See Chapter 3, Section 3-3.)

Once the transit is on-line under the wires, a sight to the pole location can be made. The field engineer is cautioned to note the type of pole and pole-to-wire connection to be used. In some cases, the pole is offset from the center line of the wires. The wires are then mounted on insulators attached to the side of the pole. If this is the case, the pole location must be offset the proper amount.

Some utility companies have the field engineer stake the center line of the wire and let the pole drilling and erecting crew allow for the offset. Other companies expect the field engineer to make allowance for the pole offset and set the stake accordingly. The field engineer must determine which method is used where he or she is working.

Some poles are placed quite close together to provide a "slack" line mounting area. This is often the case when a new line is erected perpendicular

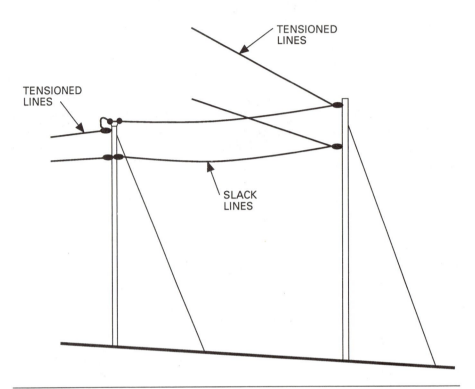

Figure 11-1. Tension and slack lines. The lines between the two poles are "slack."

to an existing line. Figure 11-1 shows a slack line area adjacent to a tensioned area.

Guy Wires

A guy wire is a diagonal brace which is designed to add strength to power lines by relieving stress on the poles. Most guy wires, if there is sufficient room, are placed at a 45° angle to the pole. At ground level, they are fastened to an anchor which is screwed into the ground. On the pole, the guy wires are fastened to specially designed fittings near the top of the pole.

The field engineer must stake the location of the guy anchors for the installation crew. This is usually done by driving a stake into the ground at the point where the guy is to contact the ground. Guy stakes are usually driven at a 45° angle and in a direction that corresponds to the installed guy. The stake is marked with the information about the guy.

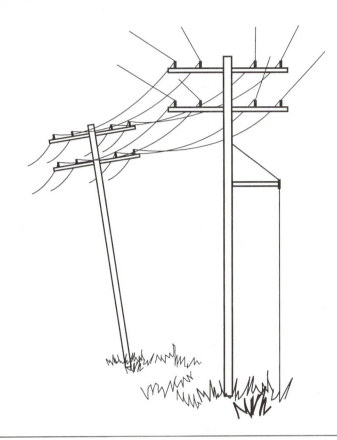

Figure 11-2.   Guy wires *not* set at a 45 degree angle to the pole.

There are some guys that are not set at 45° angles. See Figure 11-2. These guys must be properly identified so that the drilling-machine operator will install the anchor vertically or at the proper angle, rather than at a 45° angle. No elevation hubs are used when setting guy wire stakes.

11-3   Under Ground Utilities (Open Trench Construction)

Most under ground utility system construction is in the form of a pipe carrying water or other liquid or semi-liquid material. These pipelines fall into two categories, pressure and gravity flow. Both have their advantages and disadvantages as far as ease of construction goes, but both types are usually laid in open trenches.

Again, the process of laying out a pipeline is quite similar to laying out a road. The line is laid within a right-of-way and along a center line. The boundaries of the right-of-way must be identified and staked to prevent inadvertent trespassing, just as in road and overhead utility construction. The same clearing and re-staking processes apply.

The staking process includes setting indicator stakes on the center line, setting offset stakes, and setting elevation hubs. The distance offset stakes are set from the center line of the pipeline is dictated by the terrain, the depth of the trench, the slope of the sides (layback) of the trench, and by the type of excavation equipment used. Figure 11-3 shows several cross sections of pipe

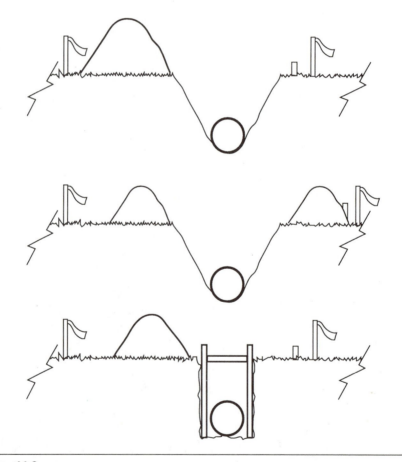

Figure 11-3.

installation projects. Note that in each case, the location of the offset stakes is dictated by varying conditions. It is the responsibility of the field engineer to consult with the project manager to determine the optimum location for these stakes.

Pressure Lines

Ditches for pressure lines are usually fairly uniform in depth and follow the existing terrain as it rises and falls. Pressure lines are usually made of steel, plastic, or fiberglass. Even the steel lines will bend enough for the pipe to follow vertical curves if the degree of change from one curve to another is not too radical.

The main responsibility of the field engineer in pressure line construction is to maintain the line of the ditch and to ensure that the depth below the existing terrain is according to the specifications. This responsibility is shared with the operator of the ditching equipment. The equipment operator must have a clear indication of the location of the center line stakes and the depth required at any point along the ditch.

Of course, additional work will be required at pump stations, valve locations, and road or river crossings. The accessory construction projects are handled the same way as any structure or bridge is handled.

Gravity Flow Lines

The ditches for gravity flow lines must be excavated to a given slope or fall. This fall must be maintained along the length of the line for proper flow to continue.

In most cases, the ditches are roughly excavated. Next, a bedding material—usually sand or fine gravel—is placed in the bottom of the ditch to adjust the elevation of the bottom of the ditch to the specified finished elevation and to provide support for each pipe joint along its entire length. This bedding material acts in much the same way as the base material used in highway construction.

It is the responsibility of the field engineer to provide line and grade to the excavation equipment operators and the pipe installers. This is accomplished by setting center line stakes, right-of-way boundary stakes, offset stakes, and elevation hubs. Batter boards may be set with string lines stretching across the

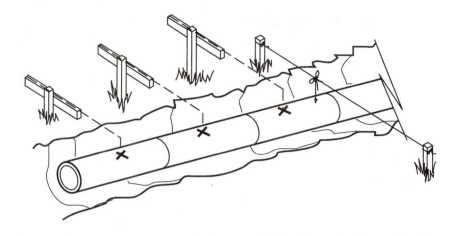

Figure 11-4. Pipeline elevation control techniques.

ditch. Figure 11-4 shows several ways to set batter boards and other elevation reference points for pipe installers.

The development of the laser level has been of particular assistance to the pipe-laying contractor. The lasers can be set in the pipe at the beginning of a run and aimed toward the open end where joints are being added. The laser beam can be adjusted to give the exact amount of rise or fall as required. Figure 11-5(a) shows a laser installed in a pipe.

A template with a sensor or a visual indicator in the center can be fitted temporarily into the open end of the joint of pipe being set. The bedding material under the pipe is then adjusted until the laser beam is striking the indicator on the template at the proper point. Once the pipe is in position, the crew moves on to the next joint and repeats the process.

Of course, if there is a bend in the line or if the laser cannot be set into the pipe for some reason, the crew must rely on the field engineer to provide grade using traditional methods. This may include the use of levels, transits, or lasers.

In most cases, an instrument is set up near the excavation and a rod is set either on the top of the pipe or on a board set in the pipe at the invert level. In the case of large-diameter pipe, the instrument may actually be set up on the pipe. If the pipe is small, the instrument may be set up in the ditch directly over the pipe. In either case, the field engineer must calculate the amount of rise or fall required for each successive joint of pipe, sight a benchmark for backsight,

(a)

(b)

Figure 11-5.   (a) A laser level inside a pipe. (Photo courtesy of Spectra Physics.) (b) A trench box in use. (Photo courtesy of Barry, Kneeland, and Schlatter and Griswold Machine & Engineering Inc.)

and sight the rod for frontsight. The mathematics for differential elevation are applied to direct the crew in adjusting the bedding to bring the pipe to the proper elevation.

Gravity flow lines made of sections of pre-cast concrete pipe are quite commonly used as storm and sanitary sewer lines. The concrete pipe is relatively inexpensive and can be made in almost any diameter. However, the pipe sections are quite heavy and must be hauled to the job site on trucks. This means they must be made in short sections, usually no more than eight feet long. Because of the short sections, a laser or other instrument must be available to indicate the proper slope of each pipe section as it is lowered into the ditch.

Another development in placing pipe in open ditches is the use of the trench box. The trench box is both a safety feature and a labor- and time-saving device. It keeps the ditch from caving in on workers, and also allows the pipe to be placed without excessive excavation having to be done. Figure 11-5(b) shows a trench box in use. Using trench boxes, the excavation contractor has only one section of the ditch open at a time. The open section is only large enough to place one joint of pipe into the ditch through the open trench box. As pipe is placed, the ditch is opened up just enough to move the trench box ahead one pipe length. The pipe joint previously laid is covered immediately as the ditch progresses. This practice means that the elevation of each pipe section must be within tolerance because it cannot be adjusted at a later time. Lasers are particularly adapted for this type of pipeline construction.

11-4   Under Ground Utilities (Tunneling)

Tunneling as a means of utility installation is almost always a last resort because of the dangers and expense involved. If tunnels are used for utilities, they are more often used for short, under-street installations where opening a trench would cause too much of a traffic obstruction in the area.

Tunneling is more often used for transportation systems such as subways or other rail systems. Often, tunnels will be used to provide rail or street crossings under existing tracks, highways, or river channels. Tunneling may be more desirable than having to construct a bridge, especially in congested areas.

Occasionally, tunnels to be used as storm and sanitary sewers are constructed. In some areas, construction and population-density patterns have changed so that existing pipelines are no longer capable of handling increased flow. Old pipes may be on the verge of collapse as well. In these congested areas, there is usually no room for deep excavations in which to install new pipes, so tunnels may be excavated, lined, and used in place of pipes.

Tunnel construction is laid out in much the same way that roads are laid out. The tunnel follows a center line and a specified grade line. Horizontal and vertical curves are also specified in much the same way as highway curves. The tunnel must remain inside a right-of-way even if it is not visible.

Some automatic tunnel-boring machines are equipped with laser systems that are used to guide the cutter heads as they grind away the material at the face of the tunnel. These machines must be kept aligned by an operator who refers to the location of the laser beam on a target located near the operator's cab or platform. If a curve or change in direction of the tunnel is required, the operator of the machine must steer the cutter heads in the proper direction. This steering is done by either offsetting the laser or the target, or by the operator keeping the target intercept point slightly off center in the proper direction while the machine advances.

Of course, the field engineer must periodically check the progress of the tunneling operation to ensure that it is on the specified line and grade, even if automatic equipment is being used.

The field engineer working on tunnel construction must overcome many obstacles that are not a problem when working on surface projects. Below is a list of some of the more serious problems in tunnel layout work and some suggestions for overcoming them.

Lack of Space to Maneuver

Tunnels by their very nature are cramped quarters. In excavating a tunnel, only the exact amount of material required to open the shaft is removed for two reasons: (1) cost and (2) engineering strength.

Of course, tunneling equipment is expensive to buy or rent and to operate. Therefore, the smallest machines capable of doing the job are used. This often results in tunnels that are quite small and hard to work in. The tunnel is also kept to the smallest diameter possible in order to preserve the strength of the surrounding material and prevent the possibility of collapse.

As a tunnel advances, the material removed from the face of the tunnel, where the actual digging is taking place, must be removed from the tunnel and deposited outside. This requires earth movers, conveyors, or some combination of machinery to move the loose material. All this machinery and its support systems, along with the ventilation system, pumps, cooling or heating lines, and communication systems quickly fill the tunnel area almost to capacity.

This leaves but one place for the field engineer to work, and that is near the roof of the tunnel. On most tunneling projects, the area near the very apex of the tunnel is reserved for sighting equipment, benchmarks, and other indicators used by the field engineer. Temporary lighting, ventilation, and other lines are usually run slightly to one side of the apex for this reason. On horizontal curves, these temporary utility lines should be run to the outside of the curve or positioned low enough on the side of the tunnel to let the field engineer sight to

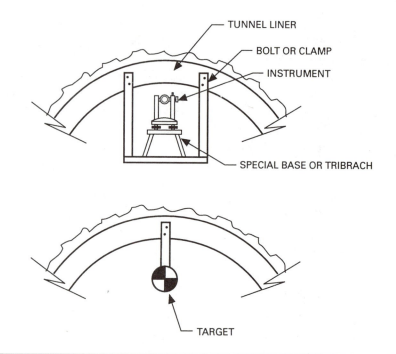

Figure 11-6. Typical installation position for an instrument and a target in a tunnel.

the inside of the curve to establish deflection points. Figure 11-6 shows one method of locating a transit and a target in a tunnel.

Vertical curves in tunnels may present yet another problem because tangent lines often run into the ceiling or floor quite quickly. Fortunately, most tunnels do not use extended vertical curves, but transition quickly to grades of varying degrees of slope and proceed as straight lines. In the case of those tunnels where extended vertical curves are required, lasers and other equipment are often used to monitor the curve as tunneling progresses. Also, some of the tunneling support equipment may have to be removed to allow the field engineers to survey the tunnel and plan corrections or continuation of the excavation process.

Short Backsights

In tunneling, one of the most severe layout problems is transferring the center line of the tunnel from the surface into the tunnel. This is particularly difficult if the tunnel is started from a shaft.

Transferring a surface line down a shaft and into a tunnel is difficult due to the absence of a long backsight on which to check alignment once the instrument is moved to the bottom of the shaft. Figure 11-7 shows the problem in profile.

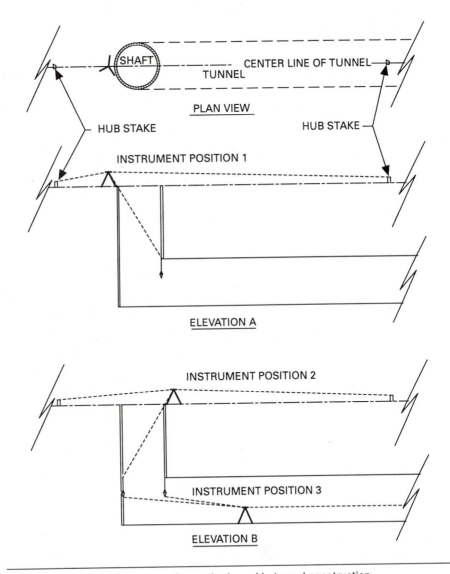

Figure 11-7. A typical line transfer method used in tunnel construction.

This problem may be overcome by using theodolites or EDMIs that are capable of reading to one second of arc, by using extreme care in setting benchmarks, and by frequently double checking the line with reference to the surface points.

Depending on the configuration of the tunnel and the shaft and the operational characteristics of the instruments being used, it may be necessary to build a small platform at ground level on each side of the shaft so that there will be a place to mount an instrument. In this case, a plumb line may be lowered from the instrument down through the floor of the platform and into the shaft. If platforms are used, an instrument with a long-focus optical plummet may be used to transfer line to the bottom of the shaft. Plumb lines lowered from batter boards may be used as well. Figure 11-8 shows these technique in profile.

When transferring grade to the tunnel area, the best method is simply to hang a tape in the shaft. Of course, the top end of the tape must be at a level related to a benchmark. The differential level between the benchmark and some point on the shaft can then be marked. Once the benchmark has been transferred to the lower shaft area, this level line can then be carried into the tunnel. Again, extreme care must be used due to the short backsight.

Dust, Water, and Heat Waves

Tunneling often presents the worst of working conditions. In addition to cramped quarters, dust is produced by the excavation equipment or blasting operations. It is often necessary to use water spray to keep the dust to a minimum. Also, water may be present in the ground where the tunnel is being dug.

The equipment used to excavate tunnels may produce quite a lot of heat. This heat has no place to dissipate, so the tunnel area may become uncomfortably warm. The surrounding ground may be noticeably warm as well.

When the combination of dust, heat, and water are encountered in the cramped quarters of a tunnel, the field engineer must take extra precautions with the surveying equipment. The equipment must be removed from the tunnel frequently and allowed to dry out. It should not be stored in the tunnel.

The dusty or hot and humid conditions in tunnels often produce haze and heat waves that interfere with the field engineer's ability to sight from point to point in the tunnel. Neither laser, EDMI, nor conventional instruments may be of any use in long sightings, so a series of short sightings must be taken. This requires greater attention to accuracy and more frequently repeated sightings.

It may be possible to do much of the field engineering work during shift

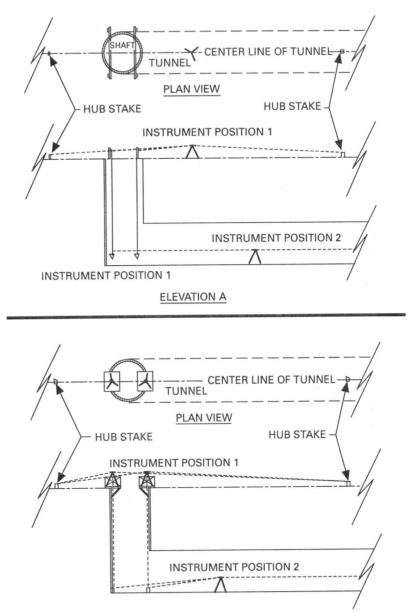

Figure 11-8.   Additional line transfer methods for tunnel construction.

changes, before work begins, on weekends, or when supplies are being moved into or out of the tunnel. Of course, this depends on crew schedules, job progress, and other factors. In any case, the visibility in the tunnel will be better after periods of inactivity and when the ventilation systems have been kept running at full capacity.

Inadequate Lighting

The lighting in most tunnels is barely adequate for the workers to perform their jobs. The light is directed to the floor of the tunnel or to the work area, so it is often quite dark at the top of the tunnel where the field engineer is trying to sight targets.

The field engineer must often have additional or supplemental lights installed and focused on the target areas. If this is not possible or is too costly, there are some alternatives. One alternative is to use battery-powered lanterns similar to those used by campers. Small platforms may be constructed of plywood or metal, the lantern set on the platform, and the light focused on the target area. These platforms may be hung from temporary bolts, clamps, or other mounting points on the tunnel liner or from anchors set in the tunnel wall. The platforms and lanterns may be easily removed, carried forward, or stored as necessary.

Deformation of Liners

Most tunnels are lined with either steel or concrete during the construction process. These liners are designed to prevent collapse of the tunnel both during construction and for years to come. While the lining material is designed to prevent collapse, it is not designed to be perfectly rigid. Some settling and deformation of the liner will occur and the roof elevation of the tunnel will decrease with time.

This settling of the liner presents a problem to the field engineer because most of the targets used for line and grade in the tunnel are attached to the liner at or near the apex. If the targets along the tunnel are settling, it will be impossible to get an accurate backsight for elevation transfer.

The rate of settlement and deformation of the liner will vary along the length of the tunnel and the rate will not be constant over time. Therefore, the field engineer must frequently run the grade line from the surface benchmarks, into the tunnel, and down the tunnel to the face of the excavation. The benchmarks along the tunnel must be adjusted or the change in their elevations must be taken into consideration mathematically as the tunnel progresses.

Surface Settlement or Subsidence

Not only does the tunnel liner change shape with time, the surface of the earth immediately over the tunnel may settle as well. This settlement is called subsidence. Most tunnel specifications will require that a surface subsidence monitoring system be installed. This monitoring system usually consists of a series of surface benchmarks set at intervals over the center line of the tunnel. Additional benchmarks may be set to either side of the center line in some areas where the soil is particularly prone to settlement or where the sub-surface soil structure is not stable.

It is the responsibility of the field engineer to install and monitor the surface subsidence system. The installation process is relatively simple. The field engineer must locate the points where monitoring is to be done, install a monument or hub, and record the elevation in relation to a known benchmark.

Periodic reports on the monitoring process will usually be required. To complete these reports, the field engineer compares the present elevation of the monitor benchmark with the elevation at the time the benchmark was installed. In some cases, a series of forms, with the elevation readings noted, must be forwarded to the proper authorities. In other cases, the only reporting that must be done is when the subsidence exceeds a certain limit. The field engineer should report the findings of the monitoring process to the project manager for action as necessary.

Due to the subsidence problem, benchmarks which are to be used for elevation control should not be set within 200 feet of the tunnel center line. This applies to benchmarks that will be used for elevation control during the construction of adjacent structures, as well as benchmarks for tunnel elevations.

Some tunneling projects must be so precise that small alignment holes are drilled from the surface down to the tunnel. This provides a check on the progress of the tunneling operation in terms of both elevation and alignment. Some of the alignment holes are enlarged to provide ventilation and other utilities to the tunneling operation.

It is often the responsibility of the field engineer to lay out the location of these alignment holes and establish reference elevation hubs with which to check the tunnel elevations. It is a matter of great pride for the field engineer and the tunneling crew when the alignment holes and the tunnel meet precisely as planned and at the proper elevation.

Tunneling Terminology

Below are some commonly used tunneling terms.

> Pitch—the vertical angle of the tunnel or vertical movement of the tunneling equipment.
>
> Yaw—the left and right angle of the tunnel or the left and right movement of the tunneling machine.
>
> Strike and dip—a term referring to the direction and angle of rock or ore formations.
>
> Shaft—usually a vertical opening into a tunneling area.
>
> Drift—a term used to refer to the tunnel itself.
>
> Portal—the entrance into the tunnel from the shaft.

Trenchless Technology

The idea of drilling small-diameter holes laterally through short distances such as under roads has been around for some time. However, drilling laterally under ground for extended distances with accuracy is a new kind of technology.

Using trenchless technology, pipes are forced through the ground, usually with the aid of high-pressure water jetting from holes in the drilling head. The shape of the drilling head and the operator's control system will allow very accurate directional control both vertically and horizontally.

Once the area to be drilled under is crossed, the drilling head is brought to the surface and removed. A reamer and a length of pipe are attached to the drill pipe and drawn back through the hole just opened by the drilling operation.

This process can be done quickly and is much less costly than opening a ditch. The disruption in the area is minimal.

Safety

Not only are tunnels cramped quarters in which to work, but there are certain dangers present. Safety briefings are mandatory prior to entering any tunnel. The project manager or the company safety representative can provide guidelines in this area.

There are some tunneling operations in which a "two-man policy" is observed. This means that no one is allowed to work in the tunnel alone. Obviously, the intent of the rule is to provide a safety measure by having each person watch out for the other.

Probably the greatest fear of people working in tunnels is that the lights will go out. Many tunnel workers use back-up lamps attached to their hard hats and there may also be battery-powered light systems available for emergencies. But, if all systems fail and all the lights in the tunnel go out, the best thing to do is to stay where you are. Other workers near the entrance will get you out in the shortest possible time. You can get hurt trying to move around in the dark.

Chapter 11 Review Questions and Exercises

1. How does the layout procedure for utility lines, tunnels, and pipelines compare to the layout procedure for roads?

2. Why is the alignment of poles in power line construction important?

3. Which type of pipeline is more critical in the elevation dimension, pressure lines or gravity flow lines?

4. List some of the major problem areas the field engineer will encounter in tunneling operations.

5. What is "trenchless technology"?

6. Observe the power and communication lines alongside the roads in your neighborhood. Note the way the guy wires are positioned, particularly on curves and corners.

12

SITE DEVELOPMENT, SPACE PLANNING, AND AS-BUILTS

12-1 Introduction

The field engineer is truly involved in a construction project from concept to move-in and beyond. This chapter will explore the field engineer's involvement from the first conceptual site-development activities through the execution of the final as-built drawings. This discussion will also include a study of space-planning activities as they might be applied to the design of new structures or existing structures which are being remodeled.

12-2 Site Development

The development of a project site is often a long and exacting process. Many individuals and organizations will be involved throughout the process. Entities involved, among others, will be environmental and neighborhood preservation groups, city planners, financial institutions, and, of course, the owner and his architect and engineers. Each individual and each organization will be trying to look out for their specialized field of interest.

Each party that gets involved in the planning of a project site will have special needs and will usually want some type of special graphic representation of the site or the surrounding area prepared for study. Someone is going to have to gather the field data for these graphics. A field engineer is often chosen for this task.

To do this type of work, the field engineer must be certain of just what type and what quantity of information is being requested and how this information is to be presented. For example, it is not unusual for the field engineer to be asked to prepare a map of an area showing streets, alleys, and other features. However, in addition to preparing the map, the field engineer may be requested to do a traffic count on the streets as well. If this is the case, the field engineer must determine such things as the time of day the count is to cover, whether the vehicles are to be classified, or, perhaps, the number of occupants in each vehicle counted.

In addition to knowing precisely which data is to be gathered, the field engineer should know how the field data is going to be treated in the office. Knowing how the data is going to be used will allow the field engineer to gather the data and report the findings in such a manner that the office personnel will be able to easily convert the field data into acceptable graphics or tables. This will save considerable time and may prevent a second trip to the field to clarify data previously gathered.

One of the most frequently used site-planning documents is the topographic map. Chapter 9 describes in detail how topographic surveys are performed and how topographic maps are developed. As in the more specialized surveys such as the traffic count mentioned above, the field engineer must know precisely what is expected in gathering the data for a topographic survey.

Of course, the main purpose of a topographic map is to show the existing elevation lines, but often additional information is requested. Probably the most frequently requested information in addition to existing elevations is the location of existing structures on the property. Another popular item frequently requested is the location of trees.

Using the request for tree location as an example, here are some of the things the field engineer might ask about prior to going to the field to gather the data.

1. The field engineer must know if all trees are to be shown on the map or only trees of a certain diameter.
2. Should the trees be divided into small, medium, or large sizes, and should the location of the trunk as well as the branch line be shown?

3. Should the species be identified?
4. Should other vegetation be shown?

This example illustrates just how complicated a request for field data might be and how precise the field engineer must be in gathering the data.

Historic Preservation

Often, a proposed development site is occupied by an older structure which many people may wish to see preserved. If this is the case, the field engineer may be required to conduct a study of the structure and record its dimensions prior to any work being done.

To conduct a study of an historic structure, the field engineer must literally start on the outside and work in. The measurements should begin with identifying an original benchmark if possible. If none are recognizable, a point on the oldest nearby structure should be referenced as a benchmark.

With the structure oriented, the exterior may be measured and the measurements recorded. Remember, the measurements should be recorded in a field book, just as is done during any field work.

The interior measurement and documentation process is carried out much like a space-planning survey. These procedures are discussed in Section 12-3.

Most areas of the country will have a local historical preservation society. These organizations assume the responsibility for the preservation of older structures and other historically significant sites. The local society will usually have information on the criteria for an historical survey. However, if the field engineer needs additional information, the U.S. Department of the Interior, National Park Service, should be consulted. This agency published guidelines for historical preservation in the Federal Register on Thursday, September 29, 1983. This document and its references have become the standard for this type of work. The Federal Register is available in most large libraries.

Procedure Note

A trick of the trade which may be used in both site development and space planning is to draw the existing plot plan or floor plan on the back of a sheet of clear mylar. Those involved in planning can then use grease pencils or erasable ink pens to sketch and plan on the top surface of the mylar without disturbing the print of the existing structure on the back. The drafter can then lay the mylar on a light-colored surface and transfer the sketches either to the back of the mylar for additional study or to another print.

To produce the existing plan on the back of the mylar, the plan must either be drawn on the top side and then transferred to the back, or it must be drawn in mirror image on the back to begin with.

As most drafting is done with computers today, this trick is even more easily done. Most drafting software will allow an image to be "mirrored." In this case, the floor plan or plot plan is mirrored and the mirrored image is plotted on the mylar sheet. Modifications specified by the planners can be transferred to the computer and the plan modified and re-plotted as necessary.

12-3   Space Planning

Interior and exterior space planning for both new and existing projects is a very important item in the construction or re-modeling process. With the costs of construction going up, optimum use of all available space is required. The field engineer is often required to assist in these activities, particularly in the remodeling of existing structures.

The space planners and interior design professionals will naturally use the prints of new structures for their planning and design efforts. However, in the case of remodeling older structures, the plans may have been lost or destroyed. Field engineers are often requested to survey the existing structures and re-create the plans.

Below are listed several problem areas normally associated with surveying older structures and some suggested solutions.

1. Hidden areas, optical illusions, and other visual anomalies may exist. Many of the cautions related to surveying historic structures apply to surveying any older structure. The walls may vary in thickness, may not run straight, and may not be plumb. There may be hidden areas where old closets or fireplace flues were covered up. In some older houses which have been remodeled numerous times, walls and offsets from the original lines may produce an "optical illusion" or other visual anomaly. These "illusions" most often occur deep within the interior of structures where there are no windows to provide an outside reference point. These visual anomalies will confuse even the most experienced field engineer, so extreme care must be taken when measuring the interiors of these structures.

Figure 12-1. Tape laid out in an old structure in preparation for gathering dimensional data.

Unless the existing structure is quite simple and easily measured, the best way to proceed is to establish base lines running from front to back and from side to side through the structure. These base lines should be established down hallways or through a series of rooms where the longest uninterrupted line can be laid out. See Figure 12-1. The distances from the base lines to walls, windows, and doors can easily be shown on a set of rough plans drawn in the field. The center line of windows and doors should also be referenced to the base line as should the center line of interior walls. These plans can easily be refined in the office and the dimensions checked.

2. Walls or other structural components may not be straight. For various reasons associated with time, the walls of older structures may be

Figure 12-2. Out-of-line condition of an old structure.

severely out of line and out of plumb. Figure 12-2 shows such a condition. The field engineer should present this information accurately and should not attempt to show the out-of-line wall in its original position. If it is out of line, it should be shown as out of line. This is another reason for using a base line and measuring laterally from it.

3. Floor plans must be drawn from measurements taken at floor level. Normally, floor plans are drawn at a plane 4 ft above floor level. This cannot be easily done when re-drawing plans for an existing structure, so the field engineer must note that measurements were taken at floor level.

When measuring at floor level, the measurement should be taken to the wall line. This means that baseboards and moldings must be accounted for. If the base moldings are not too tall, the tape can be bent up slightly to intersect the bare wall surface and the measurement taken. If the baseboards are quite tall, the measurement is taken to the surface of the baseboard at or near floor level and the thickness of the baseboard is added to the measurement. In any case, the measurement should be taken to the face of the wall. This should be noted on the plans.

Windows, cabinets, electrical switches, and other elements should be shown as if at the 4 ft plane. This will require some allowances in measurements taken at the floor level. Using a process similar to that described in measuring to the baseboards will work.

4. Wall thickness may not be uniform. The thickness of walls should be measured and noted. The standard thickness for walls has changed over the years as material standards have changed. In the 1920s, 2 × 4 lumber commonly used in wall construction actually measured 2 inches thick by 4 inches wide. Walls were covered by 1-inch thick boards. Today a 2 × 4 is 1½ inches thick by 3½ inches wide and walls are covered with ½ inch gyp-rock. Also, walls may have been recovered with various materials of varying thicknesses if the structure has been remodeled or repaired in the past. This may have produced walls of various thicknesses throughout the structure, so the actual thickness of all walls should be investigated and reported. See Figure 12-3.

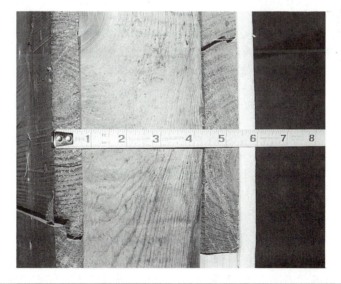

Figure 12-3. Wall assembly of an old structure showing the different layers of material and dimensions used in the original construction and in the remodeling process.

5. Walls may be out of plumb. Using a level, the field engineer should check for out-of-plumb conditions at several points in each room. Any out-of-plumb condition should be noted on the plans giving the location and the amount the wall is out of plumb.

6. The structure may not be at the same elevation as when it was built. It may have settled. More importantly, the settlement may not be uniform over the entire floor plan. This presents many problems in leveling and making components such as doors and cabinets fit properly. If possible, the field engineer should use a level and rod to determine the high and low points of the main floor of the structure. This may be done by establishing benchmarks inside the structure or it may be done by sighting through windows. In either case, the information concerning any out-of-level condition should be noted on the prints.

7. It may be difficult to determine which walls are load bearing. If it can be determined for certain which walls are load bearing and which are not, this information should be noted on the prints. Additions and remodels over the years may have made the removal of some load-bearing walls necessary. If this has been done, some other manner of supporting roof loads will have been put into place. If the field engineer is not familiar with how to determine whether a wall is load bearing, the information should not be placed on the prints until a qualified individual can check the structure.

8. The roof structure may not be safely accessible and may not be safe to walk on. If this is the case, the only thing the field engineer can do is get whatever measurements are possible using rods and ladders and report the fact to the proper authority. This also applies to porches and attics.

9. The structure may not meet the code requirements for its intended use in the jurisdiction in which it is located. If this is the case and the field engineer knows this to be a fact, the field engineer should report this information to the proper authorities. Failure to do so may result in death, serious injury, or severe property damage.

12-4 "As-Builts"

Almost no construction project is built precisely in accordance with the original plans. Changes and modifications to the original plans will have taken place during the construction process. In some cases the changes will have been so small as to be insignificant, but others may have been rather extensive.

Of course, the larger the project, the more changes in the original plans can possibly occur.

Owners and architects realize that changes are inevitable in most projects and that the original plans will not reflect perfectly the actual structure when the job is finished. For this reason, many construction contracts require that the contractor deliver a set of plans to the owner after the project has been completed which reflect all the changes which were made to the original plans. These are known as "as-built" plans or simply as "as-builts."

Most construction contracts contain very specific provisions for modifying the original plans during construction. Any change in the original plans or to the original contract usually require that a "change order" or similar document be executed. This change order describes the physical changes which are to be made and outlines the changes in the contract amount or time, or both.

The physical changes in the structure which are the result of a change order may be described by changing a portion of the original plans or by drawing a new set or partial set of plans, or the change may only be verbally described. Of course, all parties to the original contract must approve of and sign the change orders.

It is often the responsibility of the field engineer to coordinate or assist in the preparation and execution of the as-built plans. Below is a series of steps and suggestions which will assist the field engineer in completing as-built drawings.

1. Study the copy of the original plans which were used in the project office for reference. Observe and record any notes on the plans which refer to changes. Note the location and elevation of benchmarks.

2. Obtain copies of all change orders which were executed on the project. Categorize these according to the area of the plans affected. Place foundation change orders with the foundation plans, the electrical change orders with the electrical plans, and so forth.

3. Check with each subcontractor or trade foreman affected by the change orders. Do they recognize the change orders and do they recall any specifics which may differ from the description or the prints of the changed work?

4. Compare the change orders with the specifications. Again, categorize the change orders according to the specifications involved. Put change orders affecting concrete mix design with the concrete specs, and so forth.

5. Review the project manager's daily reports for reference to change orders. Review the accounts receivable for billing for change orders.

6. Review the project photographs as necessary to confirm changes.

7. Review the project manager's daily reports or diary as necessary.

8. Resolve any discrepancies between the change order documents, reports from subcontractors or foremen, billings, and any other references to changes done to the structure even if it is only a rumor. This may require visiting the structure and may even require that some work be uncovered if the changes are critical to future operations in the structure.

9. Most plans for major structures today are drawn on computers. The architect will usually provide disks with the original plans or parts of the original plans so that the drawings can be modified to reflect the as-built condition.

10. Either mark up a hard copy of the plans and specifications showing the confirmed changes or modify the computer images. In either case, a new set of plans and specifications should be produced showing the true as-built condition of the structure. Check the re-drawn plans and specifications for accuracy and submit them to the proper authorities.

As-built drawings may be kept for years and may become the primary reference used in future modifications to the structure. The as-builts may be sent to insurance agencies or fire departments to be used as references in property-protection activities. In time, the as-builts might come to have historical significance. When the possible future use of as-built drawings is realized, it then becomes a matter of pride for the field engineer to produce the best results possible.

Many owners attach such importance to the as-built documents that they preserve an original set in an atmosphere-controlled vault. Owners may even have microfilm copies made for additional security.

Chapter 12 Review Questions and Exercises

1. When asked to gather field data for site development purposes, what must the field engineer know in order to effectively carry out the assignment?

2. How can a mirror image of a site or a structure's floor plan be used in space planning?

3. When measuring a structure as a part of a space-planning assignment, should the structure be shown to be as it was or as it is?

4. What are some of the things a field engineer should be cautious of when measuring older structures?

5. List several sources of information the field engineer should consult when collecting data to be used in the production of "as-built" drawings.

6. Prepare an "as-built" floor plan of a structure. Have your instructor check your work for accuracy.

A

CONVERSION TABLES

UNIT EQUIVALENTS

1 chain	= 66 feet	= 100 links
1 link	= 0.66 feet	= 7.92 inches
1 vara	= 33.333 inches	= 2.778 feet
1 rod	= 16.5 feet	= 25 links
1 section	= 1 square mile	= 640 acres
1 acre	= 43560 square feet	= 160 square rods
1 mile (5280 feet)	= 80 chains	= 320 rods
1 labor	= 1000 square varas	
1 league	= 5000 square varas	
1 hectare	= 2.47 acres	= 1000 square meters
1 meter	= 39.37 inches	
1 kilometer	= .62 miles	
1 square kilometer	= .3861 square miles	
1 yard	= 3 feet	= 36 inches
1 furlong	= 220 yards	
1 fathom	= 6 feet	
1 nautical mile (old)	= 6080.2 feet	= 1853.2 meters
1 nautical mile (new)	= 6076.115 feet	= 1852 meters
1 inch	= 2.54 centimeters	
1 foot	= 30.4801 centimeters	
millimeters	× .04	= inches
centimeters	× .4	= inches
meters	× 3.3	= feet

meters	×	1.1	= yards
kilometers	×	.6	= miles
inches	×	2.54	= centimeters
feet	×	30.48	= centimeters
miles	×	1.6	= kilometers

CONVERTING DECIMAL DEGREES TO DEGREES, MINUTES, AND SECONDS

EXAMPLE: $26.22833° = ?$ DEGREES, ? MINUTES, ? SECONDS
$$= 26° + (.22833 \times 60')$$
$$= 26° + 13.6998'$$
$$= 26° + 13' + (.6998 \times 60'')$$
$$= 26° + 13' + 41.988''$$
$$26.22833° = 26° \ 13' \ 42''$$

CONVERTING DEGREES, MINUTES, AND SECONDS TO DECIMAL DEGREES

EXAMPLE: $26° \ 13' \ 42'' = ?$ DECIMAL DEGREES
$$= 26° + 13' + 42/60 \text{ OF A MINUTE}$$
$$= 26° + 13' + .7'$$
$$= 26° + 13.7 / 60 \text{ OF A DEGREE}°$$
$$= 26° + .22833°$$
$$26° \ 13' \ 42'' = 26.22833°$$

DECIMAL FEET TO INCHES CHART (ROUNDED)

.01'	= ⅛"	.26'	= 3⅛"	.51'	= 6⅛"	.76'	= 9⅛"
.02'	= ¼"	.27'	= 3¼"	.52'	= 6¼"	.77'	= 9¼"
03'	= ⅜"	.28'	= 3⅜"	.53'	= 6⅜"	.87'	= 9⅜"
04'	= ½"	.29'	= 3½"	.54'	= 6½"	.79'	= 9½"
.05'	= ⅝"	.30'	= 3⅝"	.55'	= 6⅝"	.80'	= 9⅝"
.06'	= ¾"	.31'	= 3¾"	.56'	6¾"	.81'	= 9¾"
.07'	= ⅞"	.32'	= 3⅞"	.57'	= 6⅞"	.82'	= 9⅞"
.08'	= 1"	.33'	= 4"	.58'	= 7"	.83'	= 10"
.09'	= 1⅛"	.34'	= 4⅛"	.59'	= 7⅛"	.84'	= 10⅛"
.10'	= 1¼"	.35'	= 4¼"	.60'	= 7¼"	.85'	= 10¼"
.11'	= 1⅜"	.36'	= 4⅜"	.61'	= 7⅜"	.86'	= 10⅜"
.12'	= 1½"	.37'	= 4½"	.62'	= 7½"	.87'	= 10½"
.13'	= 1½"	.38'	= 4½"	.63'	= 7½"	.88'	= 10½"
.14'	= 1⅝"	.39'	= 4⅝"	.64'	= 7⅝"	.89'	= 10⅝"
.15'	= 1¾"	.40'	= 4¾"	.65'	= 7¾"	.90'	= 10¾"
.16'	= 1⅞"	.41'	= 4⅞"	.66'	= 7⅞"	.91'	= 10⅞"
.17'	= 2"	.42'	= 5"	.67'	= 8"	.92'	= 11"
.18'	= 2⅛"	.43'	= 5⅛"	.68'	= 8⅛"	.93'	= 11⅛"

.19' = $2\frac{1}{4}$"	.44' = $5\frac{1}{4}$"	.69' = $8\frac{1}{4}$"	.94' = $11\frac{1}{4}$"
.20' = $2\frac{3}{8}$"	.45' = $5\frac{3}{8}$"	.70' = $8\frac{3}{8}$"	.95' = $11\frac{3}{8}$"
.21' = $2\frac{1}{2}$"	.46' = $5\frac{3}{8}$"	.71' = $8\frac{1}{2}$"	.96' = $11\frac{1}{2}$"
.22' = $2\frac{5}{8}$"	.47' = $5\frac{5}{8}$"	.72' = $8\frac{5}{8}$"	.97' = $11\frac{5}{8}$"
.23' = $2\frac{3}{4}$"	.48' = $5\frac{3}{4}$"	.73' = $8\frac{3}{4}$"	.98' = $11\frac{3}{4}$"
.24' = $2\frac{7}{8}$"	.49' = $5\frac{7}{8}$"	.74' = $8\frac{7}{8}$"	.99' = $11\frac{7}{8}$"
.25' = 3"	.50' = 6"	.75' = 9"	1.00' = 12"

CONVERTING DECIMAL FEET TO FEET AND INCHES

EXAMPLE: 28.368' = ? FEET, ? INCHES
= 28' + (12" × .368)
= 28' + 4.416"
= 28' + 4" + .416 OF ONE INCH
= 28' + 4" + (8 × .416")
= 28' + 4" + 3.352/8ths"
= 28' + $4\frac{3}{8}$" (rounded)

* Note: If the decimal feet value is to be taken to the nearest $\frac{1}{16}$" or other fraction, substitute 16 or other denominator into the formula in place of the 8 in this example.

CONVERTING FEET AND INCHES TO DECIMAL FEET

EXAMPLE: 28' $7\frac{3}{16}$" = ? DECIMAL FEET
= 28' + 7" + .1875"
= 28' + (7.1875" / 12)
= 28' + .598958'
= 28.598958'
= 28.60' (rounded)

B

TRIGONOMETRY FUNCTIONS

0.0° to 8.0

Dec/Deg	Sin	Cos	Tan	Cot	Dec/Deg
0.0	0.000 00	1.000 0	0.000 00	Infinite	90.0
.1	0.001 75	1.000 0	0.001 75	572.957	.9
.2	0.003 49	1.000 0	0.003 49	286.477	.8
.3	0.005 24	1.000 0	0.005 24	190.984	.7
.4	0.006 98	1.000 0	0.006 98	143.237	.6
.5	0.008 73	1.000 0	0.008 73	114.589	.5
.6	0.010 47	0.999 9	0.010 47	95.489	.4
.7	0.012 22	0.999 9	0.012 22	81.847	.3
.8	0.013 96	0.999 9	0.013 96	71.615	.2
.9	0.015 71	0.999 9	0.015 71	63.657	.1
1.0	0.017 45	0.999 8	0.017 46	57.290	89.0
.1	0.019 20	0.999 8	0.019 20	52.081	.9
.2	0.020 94	0.999 8	0.020 95	47.740	.8
.3	0.022 69	0.999 7	0.022 69	44.066	.7
.4	0.024 43	0.999 7	0.024 44	40.917	.6
.5	0.026 18	0.999 7	0.026 19	38.188	.5
.6	0.027 92	0.999 6	0.027 93	35.801	.4
.7	0.029 67	0.999 6	0.029 68	33.694	.3
.8	0.031 41	0.999 5	0.031 43	31.821	.2
.9	0.033 16	0.999 5	0.033 17	30.143	.1
2.0	0.034 90	0.999 4	0.034 92	28.636	88.0
.1	0.036 64	0.999 3	0.036 67	27.271	.9
.2	0.038 39	0.999 3	0.038 42	26.031	.8
.3	0.040 13	0.999 2	0.040 16	24.898	.7
.4	0.041 88	0.999 1	0.041 91	23.859	.6
.5	0.043 62	0.999 0	0.043 66	22.904	.5
.6	0.045 36	0.999 0	0.045 41	22.022	.4
.7	0.047 11	0.998 9	0.047 16	21.205	.3
.8	0.048 85	0.998 8	0.048 91	20.446	.2
.9	0.050 59	0.998 7	0.050 66	19.740	.1
3.0	0.052 34	0.998 6	0.052 41	19.081	87.0
.1	0.054 08	0.998 5	0.054 16	18.464	.9
.2	0.055 82	0.998 4	0.055 91	17.886	.8
.3	0.057 56	0.998 3	0.057 66	17.343	.7
.4	0.059 31	0.998 2	0.059 41	16.832	.6
.5	0.061 05	0.998 1	0.061 16	16.350	.5
.6	0.062 79	0.998 0	0.062 91	15.895	.4
.7	0.064 53	0.997 9	0.064 67	15.464	.3
.8	0.066 27	0.997 8	0.066 42	15.056	.2
.9	0.068 02	0.997 7	0.068 17	14.669	.1
4.0	0.069 76	0.997 6	0.069 93	14.301	86.0
.1	0.071 50	0.997 4	0.071 68	13.951	.9
.2	0.073 24	0.997 3	0.073 44	13.617	.8
.3	0.074 98	0.997 2	0.075 19	13.300	.7
.4	0.076 27	0.997 1	0.076 95	12.996	.6
.5	0.078 46	0.996 9	0.078 70	12.706	.5
.6	0.080 20	0.996 8	0.080 46	12.429	.4
.7	0.081 94	0.996 6	0.082 21	12.163	.3
.8	0.083 68	0.996 5	0.083 97	11.909	.2
.9	0.085 42	0.996 3	0.085 73	11.664	.1
5.0	0.087 16	0.996 2	0.087 49	11.430	85.0
.1	0.088 89	0.996 0	0.089 25	11.205	.9
.2	0.090 63	0.995 9	0.091 01	10.988	.8
.3	0.092 37	0.995 7	0.092 77	10.780	.7
.4	0.094 11	0.995 6	0.094 53	10.579	.6
.5	0.095 85	0.995 4	0.096 29	10.385	.5
.6	0.097 58	0.995 2	0.098 05	10.199	.4
.7	0.099 32	0.995 1	0.099 81	10.019	.3
.8	0.101 06	0.994 9	0.101 58	9.845	.2
.9	0.102 79	0.994 7	0.103 34	9.677	.1
6.0	0.104 53	0.994 5	0.105 10	9.514	84.0
.1	0.106 26	0.994 3	0.106 87	9.357	.9
.2	0.108 00	0.994 2	0.108 63	9.205	.8
.3	0.109 73	0.994 0	0.110 40	9.058	.7
.4	0.111 47	0.993 8	0.112 17	8.915	.6
.5	0.113 20	0.993 6	0.113 94	8.777	.5
.6	0.114 94	0.993 4	0.115 70	8.643	.4
.7	0.116 67	0.993 2	0.117 47	8.513	.3
.8	0.118 40	0.993 0	0.119 24	8.386	.2
.9	0.120 14	0.992 8	0.121 01	8.264	.1
7.0	0.121 87	0.992 5	0.122 78	8.144	83.0
.1	0.123 60	0.992 3	0.124 56	8.028	.9
.2	0.125 33	0.992 1	0.126 33	7.916	.8
.3	0.127 06	0.991 9	0.128 10	7.806	.7
.4	0.128 80	0.991 7	0.129 88	7.700	.6
.5	0.130 53	0.991 4	0.131 65	7.596	.5
.6	0.132 26	0.991 2	0.133 43	7.495	.4
.7	0.133 99	0.991 0	0.135 21	7.396	.3
.8	0.135 72	0.990 7	0.136 98	7.300	.2
.9	0.137 44	0.990 5	0.138 76	7.207	.1
8.0	0.139 17	0.990 3	0.140 54	7.1154	82.0
Dec/Deg	Cos	Sin	Cot	Tan	Dec/Deg

82.0 to 90.0

8.0 to 15.0

Dec/Deg	Sin	Cos	Tan	Cot	Dec/Deg
8.0	0.139 17	0.990 3	0.140 54	7.1154	82.0
.1	0.140 90	0.990 0	0.142 32	7.0264	.9
.2	0.142 63	0.989 8	0.144 10	6.9395	.8
.3	0.144 36	0.989 5	0.145 88	6.8547	.7
.4	0.146 08	0.989 3	0.147 67	6.7720	.6
.5	0.147 81	0.989 0	0.149 45	6.6912	.5
.6	0.149 54	0.988 8	0.151 24	6.6122	.4
.7	0.151 26	0.988 5	0.153 02	6.5350	.3
.8	0.152 99	0.988 2	0.154 81	6.4596	.2
.9	0.154 71	0.988 0	0.156 60	6.3859	.1
9.0	0.156 43	0.987 7	0.158 38	6.3138	81.0
.1	0.158 16	0.987 4	0.160 17	6.2432	.9
.2	0.159 88	0.987 1	0.161 96	6.1742	.8
.3	0.161 60	0.986 9	0.163 76	6.1066	.7
.4	0.163 33	0.986 6	0.165 55	6.0405	.6
.5	0.165 05	0.986 3	0.167 34	5.9758	.5
.6	0.166 77	0.986 0	0.169 14	5.9124	.4
.7	0.168 49	0.985 7	0.170 93	5.8502	.3
.8	0.170 21	0.985 4	0.172 73	5.7894	.2
.9	0.171 93	0.985 1	0.174 53	5.7297	.1
10.0	0.173 6	0.984 8	0.176 3	5.6713	80.0
.1	0.175 4	0.984 5	0.178 1	5.6140	.9
.2	0.177 1	0.984 2	0.179 9	5.5578	.8
.3	0.178 8	0.983 9	0.181 7	5.5026	.7
.4	0.180 5	0.983 6	0.183 5	5.4486	.6
.5	0.182 2	0.983 3	0.185 3	5.3955	.5
.6	0.184 0	0.982 9	0.187 1	5.3435	.4
.7	0.185 7	0.982 6	0.189 0	5.2924	.3
.8	0.187 4	0.982 3	0.190 8	5.2422	.2
.9	0.189 1	0.982 0	0.192 6	5.1929	.1
11.0	0.190 8	0.981 6	0.194 4	5.1446	79.0
.1	0.192 5	0.981 3	0.196 2	5.0970	.9
.2	0.194 2	0.981 0	0.198 0	5.0504	.8
.3	0.195 9	0.980 6	0.199 8	5.0045	.7
.4	0.197 7	0.980 3	0.201 6	4.9594	.6
.5	0.199 4	0.979 9	0.203 5	4.9152	.5
.6	0.201 1	0.979 6	0.205 3	4.8716	.4
.7	0.202 8	0.979 2	0.207 1	4.8288	.3
.8	0.204 5	0.978 9	0.208 9	4.7867	.2
.9	0.206 2	0.978 5	0.210 7	4.7453	.1
12.0	0.207 9	0.978 1	0.212 6	4.7046	78.0
.1	0.209 6	0.977 8	0.214 4	4.6646	.9
.2	0.211 3	0.977 4	0.216 2	4.6252	.8
.3	0.213 0	0.977 0	0.218 0	4.5864	.7
.4	0.214 7	0.976 7	0.219 9	4.5483	.6
.5	0.216 4	0.976 3	0.221 7	4.5107	.5
.6	0.218 1	0.975 9	0.223 5	4.4737	.4
.7	0.219 8	0.975 5	0.225 4	4.4373	.3
.8	0.221 5	0.975 1	0.227 2	4.4015	.2
.9	0.223 3	0.974 8	0.229 0	4.3662	.1
13.0	0.225 0	0.974 4	0.230 9	4.3315	77.0
.1	0.226 7	0.974 0	0.232 7	4.2972	.9
.2	0.228 4	0.973 6	0.234 5	4.2635	.8
.3	0.230 0	0.973 2	0.236 4	4.2303	.7
.4	0.231 7	0.972 8	0.238 2	4.1976	.6
.5	0.233 4	0.972 4	0.240 1	4.1653	.5
.6	0.235 1	0.972 0	0.241 9	4.1335	.4
.7	0.236 8	0.971 5	0.243 8	4.1022	.3
.8	0.238 5	0.971 1	0.245 6	4.0713	.2
.9	0.240 2	0.970 7	0.247 5	4.0408	.1
14.0	0.241 9	0.970 3	0.249 3	4.0108	76.0
.1	0.243 6	0.969 9	0.251 2	3.9812	.9
.2	0.245 3	0.969 4	0.253 0	3.9520	.8
.3	0.247 0	0.969 0	0.254 9	3.9232	.7
.4	0.248 7	0.968 6	0.256 8	3.8947	.6
.5	0.250 4	0.968 1	0.258 6	3.8667	.5
.6	0.252 1	0.967 7	0.260 5	3.8391	.4
.7	0.253 8	0.967 3	0.262 3	3.8118	.3
.8	0.255 4	0.966 8	0.264 2	3.7843	.2
.9	0.257 1	0.966 4	0.266 1	3.7583	.1
15.0	0.258 8	0.965 9	0.267 9	3.7321	75.0
.1	0.260 5	0.965 5	0.269 8	3.7062	.9
.2	0.262 2	0.965 0	0.271 7	3.6806	.8
.3	0.263 9	0.964 6	0.273 6	3.6554	.7
.4	0.265 6	0.964 1	0.275 4	3.6305	.6
.5	0.267 2	0.963 6	0.277 3	3.6059	.5
.6	0.268 9	0.963 2	0.279 2	3.5816	.4
.7	0.270 6	0.962 7	0.281 1	3.5576	.3
.8	0.272 3	0.962 2	0.283 0	3.5339	.2
.9	0.274 0	0.961 7	0.284 9	3.5105	.1
16.0	0.275 6	0.961 3	0.286 7	3.4874	74.0
Dec/Deg	Cos	Sin	Cot	Tan	Dec/Deg

74.0 to 82.0

(continued)

16.0 to 24.0

Dec/Deg	Sin	Cos	Tan	Cot	Dec/Deg
16.0	0.275 6	0.961 3	0.286 7	3.4874	74.0
.1	0.277 3	0.960 8	0.288 6	3.4646	.9
.2	0.279 0	0.960 3	0.290 5	3.4420	.8
.3	0.280 7	0.959 8	0.292 4	3.4197	.7
.4	0.282 3	0.959 3	0.294 3	3.3977	.6
.5	0.284 0	0.958 8	0.296 2	3.3759	.5
.6	0.285 7	0.958 3	0.298 1	3.3544	.4
.7	0.287 4	0.957 8	0.300 0	3.3332	.3
.8	0.289 0	0.957 3	0.301 9	3.3122	.2
.9	0.290 7	0.956 8	0.303 8	3.2914	.1
17.0	0.292 4	0.956 3	0.305 7	3.2709	73.0
.1	0.294 0	0.955 8	0.307 6	3.2506	.9
.2	0.295 7	0.955 3	0.309 6	3.2305	.8
.3	0.297 4	0.954 8	0.311 5	3.2106	.7
.4	0.299 0	0.954 2	0.313 4	3.1910	.6
.5	0.300 7	0.953 7	0.315 3	3.1716	.5
.6	0.302 4	0.953 2	0.317 2	3.1524	.4
.7	0.304 0	0.952 7	0.319 1	3.1334	.3
.8	0.305 7	0.952 1	0.321 1	3.1146	.2
.9	0.307 4	0.951 6	0.323 0	3.0961	.1
18.0	0.309 0	0.951 1	0.324 9	3.0777	72.0
.1	0.310 7	0.950 5	0.326 9	3.0595	.9
.2	0.312 3	0.950 0	0.328 8	3.0415	.8
.3	0.314 0	0.949 4	0.330 7	3.0237	.7
.4	0.315 6	0.948 9	0.332 7	3.0061	.6
.5	0.317 3	0.948 3	0.334 6	2.9887	.5
.6	0.319 0	0.947 8	0.336 5	2.9714	.4
.7	0.320 6	0.947 2	0.338 5	2.9544	.3
.8	0.322 3	0.946 6	0.340 4	2.9375	.2
.9	0.323 9	0.946 1	0.342 4	2.9208	.1
19.0	0.325 6	0.945 5	0.344 3	2.9042	71.0
.1	0.327 2	0.944 9	0.346 3	2.8878	.9
.2	0.328 9	0.944 4	0.348 2	2.8716	.8
.3	0.330 5	0.943 8	0.350 2	2.8556	.7
.4	0.332 2	0.943 2	0.352 2	2.8397	.6
.5	0.333 8	0.942 6	0.354 1	2.8239	.5
.6	0.335 5	0.942 1	0.356 1	2.8083	.4
.7	0.337 1	0.941 5	0.358 1	2.7929	.3
.8	0.338 7	0.940 9	0.360 0	2.7776	.2
.9	0.340 4	0.940 3	0.362 0	2.7625	.1
20.0	0.342 0	0.939 7	0.364 0	2.7475	70.0
.1	0.343 7	0.939 1	0.365 9	2.7326	.9
.2	0.345 3	0.938 5	0.367 9	2.7179	.8
.3	0.346 9	0.937 9	0.369 9	2.7034	.7
.4	0.348 6	0.937 3	0.371 9	2.6889	.6
.5	0.350 2	0.936 7	0.373 9	2.6746	.5
.6	0.351 8	0.936 1	0.375 9	2.6605	.4
.7	0.353 5	0.935 4	0.377 9	2.6464	.3
.8	0.355 1	0.934 8	0.379 9	2.6325	.2
.9	0.356 7	0.934 2	0.381 9	2.6187	.1
21.0	0.358 4	0.933 6	0.383 9	2.6051	69.0
.1	0.360 0	0.933 0	0.385 9	2.5916	.9
.2	0.361 6	0.932 3	0.387 9	2.5782	.8
.3	0.363 3	0.931 7	0.389 9	2.5649	.7
.4	0.364 9	0.931 1	0.391 9	2.5517	.6
.5	0.366 5	0.930 4	0.393 9	2.5386	.5
.6	0.368 1	0.929 8	0.395 9	2.5257	.4
.7	0.369 7	0.929 1	0.397 9	2.5129	.3
.8	0.371 4	0.928 5	0.400 0	2.5002	.2
.9	0.373 0	0.927 8	0.402 0	2.4876	.1
22.0	0.374 6	0.927 2	0.404 0	2.4751	68.0
.1	0.376 5	0.926 5	0.406 1	2.4627	.9
.2	0.377 8	0.925 9	0.408 1	2.4504	.8
.3	0.379 5	0.925 2	0.410 1	2.4383	.7
.4	0.381 1	0.924 5	0.412 2	2.4262	.6
.5	0.382 7	0.923 9	0.414 2	2.4142	.5
.6	0.384 3	0.923 2	0.416 3	2.4021	.4
.7	0.385 9	0.922 5	0.418 3	2.3906	.3
.8	0.387 5	0.921 9	0.420 4	2.3789	.2
.9	0.389 1	0.921 2	0.422 4	2.3673	.1
23.0	0.390 7	0.920 5	0.424 5	2.3559	67.0
.1	0.392 3	0.919 8	0.426 5	2.3445	.9
.2	0.393 9	0.919 1	0.428 6	2.3332	.8
.3	0.395 5	0.918 4	0.430 7	2.3220	.7
.4	0.397 1	0.917 8	0.432 7	2.3109	.6
.5	0.398 7	0.917 1	0.434 8	2.2998	.5
.6	0.400 3	0.916 4	0.436 9	2.2889	.4
.7	0.401 9	0.915 7	0.439 0	2.2781	.3
.8	0.403 5	0.915 0	0.441 1	2.2673	.2
.9	0.405 1	0.914 3	0.443 1	2.2566	.1
24.0	0.406 7	0.913 5	0.445 2	2.2460	66.0

Dec/Deg	Cos	Sin	Cot	Tan	Dec/Deg

66.0 to 74.0

24.0° to 32.0°

Dec/Deg	Sin	Cos	Tan	Cot	Dec/Deg
24.0	0.406 7	0.913 5	0.445 2	2.2460	66.0
.1	0.408 3	0.912 8	0.447 3	2.2355	.9
.2	0.409 9	0.912 1	0.449 4	2.2251	.8
.3	0.411 5	0.911 4	0.451 5	2.2148	.7
.4	0.413 1	0.910 7	0.453 6	2.2045	.6
.5	0.414 7	0.910 0	0.455 7	2.1943	.5
.6	0.416 3	0.909 2	0.457 8	2.1842	.4
.7	0.417 9	0.908 5	0.459 9	2.1742	.3
.8	0.419 5	0.907 8	0.462 1	2.1642	.2
.9	0.421 0	0.907 0	0.464 2	2.1543	.1
25.0	0.422 6	0.906 3	0.466 3	2.1445	65.0
.1	0.424 2	0.905 6	0.468 4	2.1348	.9
.2	0.425 8	0.904 8	0.470 6	2.1251	.8
.3	0.427 4	0.904 1	0.472 7	2.1155	.7
.4	0.428 9	0.903 3	0.474 8	2.1060	.6
.5	0.430 5	0.902 6	0.477 0	2.0965	.5
.6	0.432 1	0.901 8	0.479 1	2.0872	.4
.7	0.433 7	0.901 1	0.481 3	2.0778	.3
.8	0.435 2	0.900 3	0.483 4	2.0686	.2
.9	0.436 8	0.899 6	0.485 6	2.0594	.1
26.0	0.438 4	0.898 8	0.487 7	2.0503	64.0
.1	0.439 9	0.898 0	0.489 9	2.0413	.9
.2	0.441 5	0.897 3	0.492 1	2.0323	.8
.3	0.443 1	0.896 5	0.494 2	2.0233	.7
.4	0.444 6	0.895 7	0.496 4	2.0145	.6
.5	0.446 2	0.894 9	0.498 6	2.0057	.5
.6	0.447 8	0.894 2	0.500 8	1.9970	.4
.7	0.449 3	0.893 4	0.502 9	1.9883	.3
.8	0.450 9	0.892 6	0.505 1	1.9797	.2
.9	0.452 4	0.891 8	0.507 3	1.9711	.1
27.0	0.454 0	0.891 0	0.509 5	1.9626	63.0
.1	0.455 5	0.890 2	0.511 7	1.9542	.9
.2	0.457 1	0.889 4	0.513 9	1.9458	.8
.3	0.458 6	0.888 6	0.516 1	1.9375	.7
.4	0.460 2	0.887 8	0.518 4	1.9292	.6
.5	0.461 7	0.887 0	0.520 6	1.9210	.5
.6	0.463 3	0.886 2	0.522 8	1.9128	.4
.7	0.464 8	0.885 4	0.525 0	1.9047	.3
.8	0.466 4	0.884 6	0.527 2	1.8967	.2
.9	0.467 9	0.883 8	0.529 5	1.8867	.1
28.0	0.469 5	0.882 9	0.531 7	1.8807	62.0
.1	0.471 0	0.882 1	0.534 0	1.8728	.9
.2	0.472 6	0.881 3	0.536 2	1.8650	.8
.3	0.474 1	0.880 5	0.538 4	1.8572	.7
.4	0.475 6	0.879 6	0.540 7	1.8495	.6
.5	0.477 2	0.878 8	0.543 0	1.8418	.5
.6	0.478 7	0.878 0	0.545 2	1.8341	.4
.7	0.480 2	0.877 1	0.547 5	1.8265	.3
.8	0.481 8	0.876 3	0.549 8	1.8190	.2
.9	0.483 3	0.875 5	0.552 0	1.8115	.1
29.0	0.484 8	0.874 6	0.554 3	1.8040	61.0
.1	0.486 3	0.873 8	0.556 6	1.7966	.9
.2	0.487 9	0.872 9	0.558 9	1.7893	.8
.3	0.489 4	0.872 1	0.561 2	1.7820	.7
.4	0.490 9	0.871 2	0.563 5	1.7747	.6
.5	0.492 4	0.870 4	0.565 8	1.7675	.5
.6	0.493 9	0.869 5	0.568 1	1.7603	.4
.7	0.495 5	0.868 6	0.570 4	1.7532	.3
.8	0.497 0	0.867 8	0.572 7	1.7461	.2
.9	0.498 5	0.866 9	0.575 0	1.7391	.1
30.0	0.500 0	0.866 0	0.577 4	1.7321	60.0
.1	0.501 5	0.865 2	0.579 7	1.7251	.9
.2	0.503 0	0.864 3	0.582 0	1.7182	.8
.3	0.504 5	0.863 4	0.584 4	1.7113	.7
.4	0.506 0	0.862 5	0.586 7	1.7045	.6
.5	0.507 5	0.861 6	0.589 0	1.6977	.5
.6	0.509 0	0.860 7	0.591 4	1.6909	.4
.7	0.510 5	0.859 9	0.593 8	1.6842	.3
.8	0.512 0	0.859 0	0.596 1	1.6775	.2
.9	0.513 5	0.858 1	0.598 5	1.6709	.1
31.0	0.515 0	0.857 2	0.600 9	1.6643	59.0
.1	0.516 5	0.856 3	0.603 2	1.6577	.9
.2	0.518 0	0.855 4	0.605 6	1.6512	.8
.3	0.519 5	0.854 5	0.608 0	1.6447	.7
.4	0.521 0	0.853 6	0.610 4	1.6383	.6
.5	0.522 5	0.852 6	0.612 8	1.6319	.5
.6	0.524 0	0.851 7	0.615 2	1.6255	.4
.7	0.525 5	0.850 8	0.617 6	1.6191	.3
.8	0.527 0	0.849 9	0.620 0	1.6128	.2
.9	0.528 4	0.849 0	0.622 4	1.6066	.1
32.0	0.529 9	0.848 0	0.624 9	1.6003	58.0

Dec/Deg	Cos	Sin	Cot	Tan	Dec/Deg

58.0° to 66.0°

32.0° to 39.0°

Dec/Deg	Sin	Cos	Tan	Cot	Dec/Deg
32.0	0.5299	0.8480	0.6249	1.6003	58.0
.1	0.5314	0.8471	0.6273	1.5941	.9
.2	0.5329	0.8462	0.6297	1.5880	.8
.3	0.5344	0.8453	0.6322	1.5818	.7
.4	0.5358	0.8443	0.6346	1.5757	.6
.5	0.5373	0.8434	0.6371	1.5697	.5
.6	0.5388	0.8425	0.6395	1.5637	.4
.7	0.5402	0.8415	0.6420	1.5577	.3
.8	0.5417	0.8406	0.6445	1.5517	.2
.9	0.5432	0.8396	0.6469	1.5458	.1
33.0	0.5446	0.8387	0.6494	1.5399	57.0
.1	0.5461	0.8377	0.6519	1.5340	.9
.2	0.5476	0.8368	0.6544	1.5282	.8
.3	0.5490	0.8358	0.6569	1.5224	.7
.4	0.5505	0.8348	0.6594	1.5166	.6
.5	0.5519	0.8339	0.6619	1.5108	.5
.6	0.5534	0.8329	0.6644	1.5051	.4
.7	0.5548	0.8320	0.6669	1.4994	.3
.8	0.5563	0.8310	0.6694	1.4938	.2
.9	0.5577	0.8300	0.6720	1.4882	.1
34.0	0.5592	0.8290	0.6745	1.4826	56.0
.1	0.5606	0.8281	0.6771	1.4770	.9
.2	0.5621	0.8271	0.6796	1.4715	.8
.3	0.5635	0.8261	0.6822	1.4659	.7
.4	0.5650	0.8251	0.6847	1.4605	.6
.5	0.5664	0.8241	0.6873	1.4550	.5
.6	0.5678	0.8231	0.6899	1.4496	.4
.7	0.5693	0.8221	0.6924	1.4442	.3
.8	0.5707	0.8211	0.6950	1.4388	.2
.9	0.5721	0.8202	0.6976	1.4335	.1
35.0	0..5736	0.8192	0.7002	1.4281	55.0
.1	0.5750	0.8181	0.7028	1.4229	.9
.2	0.5764	0.8171	0.7054	1.4176	.8
.3	0.5779	0.8161	0.7080	1.4124	.7
.4	0.5793	0.8151	0.7107	1.4071	.6
.5	0.5807	0.8141	0.7133	1.4019	.5
.6	0.5821	0.8131	0.7159	1.3968	.4
.7	0.5835	0.8121	0.7186	1.3916	.3
.8	0.5850	0.8111	0.7212	1.3865	.2
.9	0.5864	0.8100	0.7239	1.3814	.1
36.0	0.5878	0.8090	0.7265	1.3764	54.0
.1	0.5892	0.8080	0.7292	1.3713	.9
.2	0.5906	0.8070	0.7319	1.3663	.8
.3	0.5920	0.8059	0.7346	1.3613	.7
.4	0.5934	0.8049	0.7373	1.3564	.6
.5	0.5948	0.8039	0.7400	1.3514	.5
.6	0.5962	0.8028	0.7427	1.3465	.4
.7	0.5976	0.8018	0.7454	1.3416	.3
.8	0.5990	0.8007	0.7481	1.3367	.2
.9	0.6004	0.7997	0.7508	1.3319	.1
37.0	0.6018	0.7986	0.7536	1.3270	53.0
.1	0.6032	0.7976	0.7563	1.3222	.9
.2	0.6046	0.7965	0.7590	1.3175	.8
.3	0.6060	0.7955	0.7618	1.3127	.7
.4	0.6074	0.7944	0.7646	1.3079	.6
.5	0.6088	0.7934	0.7673	1.3032	.5
.6	0.6101	0.7923	0.7701	1.2985	.4
.7	0.6115	0.7912	0.7729	1.2938	.3
.8	0.6129	0.7902	0.7757	1.2892	.2
.9	0.6143	0.7891	0.7785	1.2846	.1
38.0	0.6157	0.7880	0.7813	1.2799	52.0
.1	0.6170	0.7869	0.7841	1.2753	.9
.2	0.6184	0.7859	0.7869	1.2708	.8
.3	0.6198	0.7848	0.7898	1.2662	.7
.4	0.6211	0.7837	0.7926	1.2617	.6
.5	0.6225	0.7826	0.7954	1.2572	.5
.6	0.6239	0.7815	0.7983	1.2527	.4
.7	0.6252	0.7804	0.8012	1.2482	.3
.8	0.6266	0.7793	0.8040	1.2437	.2
.9	0.6280	0.7782	0.8069	1.2393	.1
39.0	0.6293	0.7771	0.8098	1.2349	51.0

Dec/Deg	Cos	Sin	Cot	Tan	Dec/Deg

51.0° to 58.0°

39.0° to 45.0°

Dec/Deg	Sin	Cos	Tan	Cot	Dec/Deg
39.0	0.6293	0.7771	0.8098	1.2349	51.0
.1	0.6307	0.7760	0.8127	1.2305	.9
.2	0.6320	0.7749	0.8156	1.2261	.8
.3	0.6334	0.7738	0.8185	1.2218	.7
.4	0.6347	0.7727	0.8214	1.2174	.6
.5	0.6361	0.7716	0.8243	1.2131	.5
.6	0.6374	0.7705	0.8273	1.2088	.4
.7	0.6388	0.7694	0.8302	1.2045	.3
.8	0.6401	0.7683	0.8332	1.2002	.2
.9	0.6414	0.7672	0.8361	1.1960	.1
40.0	0.6428	0.7660	0.8391	1.1918	50.0
.1	0.6441	0.7649	0.8421	1.1875	.9
.2	0.6455	0.7638	0.8451	1.1833	.8
.3	0.6468	0.7627	0.8481	1.1792	.7
.4	0.6481	0.7615	0.8511	1.1750	.6
.5	0.6494	0.7604	0.8541	1.1708	.5
.6	0.6508	0.7593	0.8571	1.1667	.4
.7	0.6521	0.7581	0.8601	1.1626	.3
.8	0.6534	0.7570	0.8632	1.1585	.2
.9	0.6547	0.7559	0.8662	1.1544	.1
41.0	0.6561	0.7547	0.8693	1.1504	49.0
.1	0.6574	0.7536	0.8724	1.1463	.9
.2	0.6587	0.7524	0.8754	1.1423	.8
.3	0.6600	0.7513	0.8785	1.1383	.7
.4	0.6613	0.7501	0.8816	1.1343	.6
.5	0.6626	0.7490	0.8847	1.1303	.5
.6	0.6639	0.7478	0.8878	1.1263	.4
.7	0.6652	0.7466	0.8910	1.1224	.3
.8	0.6665	0.7455	0.8941	1.1185	.2
.9	0.6678	0.7443	0.8972	1.1145	.1
42.0	0.6691	0.7431	0.9004	1.1106	48.0
.1	0.6704	0.7420	0.9036	1.1067	.9
.2	0.6717	0.7408	0.9067	1.1028	.8
.3	0.6730	0.7396	0.9099	1.0990	.7
.4	0.6743	0.7385	0.9131	1.0951	.6
.5	0.6756	0.7373	0.9163	1.0913	.5
.6	0.6769	0.7361	0.9195	1.0875	.4
.7	0.6782	0.7349	0.9228	1.0837	.3
.8	0.6794	0.7337	0.9260	1.0799	.2
.9	0.6807	0.7325	0.9293	1.0761	.1
43.0	0.6820	0.7314	0.9325	1.0724	47.0
.1	0.6833	0.7302	0.9358	1.0686	.9
.2	0.6845	0.7290	0.9391	1.0649	.8
.3	0.6858	0.7278	0.9424	1.0612	.7
.4	0.6871	0.7266	0.9457	1.0575	.6
.5	0.6884	0.7254	0.9490	1.0538	.5
.6	0.6896	0.7242	0.9523	1.0501	.4
.7	0.6909	0.7230	0.9556	1.0464	.3
.8	0.6921	0.7218	0.9590	1.0428	.2
.9	0.6934	0.7206	0.9623	1.0392	.1
44.0	0.6947	0.7193	0.9657	1.0355	46.0
.1	0.6959	0.7181	0.9691	1.0319	.9
.2	0.6972	0.7169	0.9725	1.0283	.8
.3	0.6984	0.7157	0.9759	1.0247	.7
.4	0.6997	0.7145	0.9793	1.0212	.6
.5	0.7009	0.7133	0.9827	1.0176	.5
.6	0.7022	0.7120	0.9861	1.0141	.4
.7	0.7034	0.7108	0.9896	1.0105	.3
.8	0.7046	0.7096	0.9930	1.0070	.2
.9	0.7059	0.7083	0.9965	1.0035	.1
45.0	0.7071	0.7071	1.0000	1.0000	45.0

Dec/Deg	Cos	Sin	Cot	Tan	Dec/Deg

45.0° to 51.0°

Minutes	0°		1°		2°		3°	
	Hor. dist.	Diff. elev.	Hor. dist.	Diff. elev.	Hor. dist.	Diff. elev.	Hor. dist.	Diff. elev.
0	100.00	0.00	99.97	1.74	99.88	3.49	99.73	5.23
2	100.00	0.06	99.97	1.80	99.87	3.55	99.72	5.28
4	100.00	0.12	99.97	1.86	99.87	3.60	99.71	5.34
6	100.00	0.17	99.96	1.92	99.87	3.66	99.71	5.40
8	100.00	0.23	99.96	1.98	99.86	3.72	99.70	5.46
10	100.00	0.29	99.96	2.04	99.86	3.78	99.69	5.52
12	100.00	0.35	99.96	2.09	99.85	3.84	99.69	5.57
14	100.00	0.41	99.95	2.15	99.85	3.90	99.68	5.63
16	100.00	0.47	99.95	2.21	99.84	3.95	99.68	5.69
18	100.00	0.52	99.95	2.27	99.84	4.01	99.67	5.75
20	100.00	0.58	99.95	2.33	99.83	4.07	99.66	5.80
22	100.00	0.64	99.94	2.38	99.83	4.13	99.66	5.86
24	100.00	0.70	99.94	2.44	99.82	4.18	99.65	5.92
26	99.99	0.76	99.94	2.50	99.82	4.24	99.64	5.98
28	99.99	0.81	99.93	2.56	99.81	4.30	99.63	6.04
30	99.99	0.87	99.93	2.62	99.81	4.36	99.63	6.09
32	99.99	0.93	99.93	2.67	99.80	4.42	99.62	6.15
34	99.99	0.99	99.93	2.73	99.80	4.48	99.62	6.21
36	99.99	1.05	99.92	2.79	99.79	4.53	99.61	6.27
38	99.99	1.11	99.92	2.85	99.79	4.59	99.60	6.33
40	99.99	1.16	99.92	2.91	99.78	4.65	99.59	6.38
42	99.99	1.22	99.91	2.97	99.78	4.71	99.59	6.44
44	99.98	1.28	99.91	3.02	99.77	4.76	99.58	6.50
46	99.98	1.34	99.90	3.08	99.77	4.82	99.57	6.56
48	99.98	1.40	99.90	3.14	99.76	4.88	99.56	6.61
50	99.98	1.45	99.90	3.20	99.76	4.94	99.56	6.67
52	99.98	1.51	99.89	3.26	99.75	4.99	99.55	6.73
54	99.98	1.57	99.89	3.31	99.74	5.05	99.54	6.78
56	99.97	1.63	99.89	3.37	99.74	5.11	99.53	6.84
58	99.97	1.69	99.88	3.43	99.73	5.17	99.52	6.90
60	99.97	1.74	99.88	3.49	99.73	5.23	99.51	6.96
C = 0.75	0.75	0.01	0.75	0.02	0.75	0.03	0.75	0.05
C = 1.00	1.00	0.01	1.00	0.03	1.00	0.04	1.00	0.06
C = 1.25	1.25	0.02	1.25	0.03	1.25	0.05	1.25	0.08

Minutes	4°		5°		6°		7°	
	Hor. dist.	Diff. elev.	Hor. dist.	Diff. elev.	Hor. dist.	Diff. elev.	Hor. dist.	Diff. elev.
0	99.51	6.96	99.24	8.68	98.91	10.40	98.51	12.10
2	99.51	7.02	99.23	8.74	98.90	10.45	98.50	12.15
4	99.50	7.07	99.22	8.80	98.88	10.51	98.48	12.21
6	99.49	7.13	99.21	8.85	98.87	10.57	98.47	12.26
8	99.48	7.19	99.20	8.91	98.86	10.62	98.46	12.32
10	99.47	7.25	99.19	8.97	98.85	10.68	98.44	12.38
12	99.46	7.30	99.18	9.03	98.83	10.74	98.43	12.43
14	99.46	7.36	99.17	9.08	98.82	10.79	98.41	12.49
16	99.45	7.42	99.16	9.14	98.81	10.85	98.40	12.55
18	99.44	7.48	99.15	9.20	98.80	10.91	98.39	12.60
20	99.43	7.53	99.14	9.25	98.78	10.96	98.37	12.66
22	99.42	7.59	99.13	9.31	98.77	11.02	98.36	12.72
24	99.41	7.65	99.11	9.37	98.76	11.08	98.34	12.77
26	99.40	7.71	99.10	9.43	98.74	11.13	98.33	12.83
28	99.39	7.76	99.09	9.48	98.73	11.19	98.31	12.88
30	99.38	7.82	99.08	9.54	98.72	11.25	98.29	12.94
32	99.38	7.88	99.07	9.60	98.71	11.30	98.28	13.00
34	99.37	7.94	99.06	9.65	98.69	11.36	98.27	13.05
36	99.36	7.99	99.05	9.71	98.68	11.42	98.25	13.11
38	99.35	8.05	99.04	9.77	98.67	11.47	98.24	13.17
40	99.34	8.11	99.03	9.83	98.65	11.53	98.22	13.22
42	99.33	8.17	99.01	9.88	98.64	11.59	98.20	13.28
44	99.32	8.22	99.00	9.94	98.63	11.64	98.19	13.33
46	99.31	8.28	98.99	10.00	98.61	11.70	98.17	13.39
48	99.30	8.34	98.98	10.05	98.60	11.76	98.16	13.45
50	99.29	8.40	98.97	10.11	98.58	11.81	98.14	13.50
52	99.28	8.45	98.96	10.17	98.57	11.87	98.13	13.56
54	99.27	8.51	98.94	10.22	98.56	11.93	98.11	13.61
56	99.26	8.57	98.93	10.28	98.54	11.98	98.10	13.67
58	99.25	8.63	98.92	10.34	98.53	12.04	98.08	13.73
60	99.24	8.68	98.91	10.40	98.51	12.10	98.06	13.78
C=0.75	0.75	0.06	0.75	0.07	0.75	0.08	0.74	0.10
C=1.00	1.00	0.08	0.99	0.09	0.99	0.11	0.99	0.13
C=1.25	1.25	0.10	1.24	0.11	1.24	0.14	1.24	0.16

Minutes	8° Hor. dist.	8° Diff. elev.	9° Hor. dist.	9° Diff. elev.	10° Hor. dist.	10° Diff. elev.	11° Hor. dist.	11° Diff. elev.
0	98.06	13.78	97.55	15.45	96.98	17.10	96.36	18.73
2	98.05	13.84	97.53	15.51	96.96	17.16	96.34	18.78
4	98.03	13.89	97.52	15.56	96.94	17.21	96.32	18.84
6	98.01	13.95	97.50	15.62	96.92	17.26	96.29	18.89
8	98.00	14.01	97.48	15.67	96.90	17.32	96.27	18.95
10	97.98	14.06	97.46	15.73	96.88	17.37	96.25	19.00
12	97.97	14.12	97.44	15.78	96.86	17.43	96.23	19.05
14	97.95	14.17	97.43	15.84	96.84	17.48	96.21	19.11
16	97.93	14.23	97.41	15.89	96.82	17.54	96.18	19.16
18	97.92	14.28	97.39	15.95	96.80	17.59	96.16	19.21
20	97.90	14.34	97.37	16.00	96.78	17.65	96.14	19.27
22	97.88	14.40	97.35	16.06	96.76	17.70	96.12	19.32
24	97.87	14.45	97.33	16.11	96.74	17.76	96.09	19.38
26	97.85	14.51	97.31	16.17	96.72	17.81	96.07	19.43
28	97.83	14.56	97.29	16.22	96.70	17.86	96.05	19.48
30	97.82	14.62	97.28	16.28	96.68	17.92	96.03	19.54
32	97.80	14.67	97.26	16.33	96.66	17.97	96.00	19.59
34	97.78	14.73	97.24	16.39	96.64	18.03	95.98	19.64
36	97.76	14.79	97.22	16.44	96.62	18.08	95.96	19.70
38	97.75	14.84	97.20	16.50	96.60	18.14	95.93	19.75
40	97.73	14.90	97.18	16.55	96.57	18.19	95.91	19.80
42	97.71	14.95	97.16	16.61	96.55	18.24	95.89	19.86
44	97.69	15.01	97.14	16.66	96.53	18.30	95.86	19.91
46	97.68	15.06	97.12	16.72	96.51	18.35	95.84	19.96
48	97.66	15.12	97.10	16.77	96.49	18.41	95.82	20.02
50	97.64	15.17	97.08	16.83	96.47	18.46	95.79	20.07
52	97.62	15.23	97.06	16.88	96.45	18.51	95.77	20.12
54	97.61	15.28	97.04	16.94	96.42	18.57	95.75	20.18
56	97.59	15.34	97.02	16.99	96.40	18.62	95.72	20.23
58	97.57	15.40	97.00	17.05	96.38	18.68	95.70	20.28
60	97.55	15.45	96.98	17.10	96.36	18.73	95.68	20.34
C = 0.75	0.74	0.11	0.74	0.12	0.74	0.14	0.73	0.15
C = 1.00	0.99	0.15	0.99	0.16	0.98	0.18	0.98	0.20
C = 1.25	1.23	0.18	1.23	0.21	1.23	0.23	1.22	0.25

Minutes	12°		13°		14°		15°	
	Hor. dist.	Diff. elev.	Hor. dist.	Diff. elev.	Hor. dist.	Diff. elev.	Hor. dist.	Diff. elev.
0	95.68	20.34	94.94	21.92	94.15	23.47	93.30	25.00
2	95.65	20.39	94.91	21.97	94.12	23.52	93.27	25.05
4	95.63	20.44	94.89	22.02	94.09	23.58	93.24	25.10
6	95.61	20.50	94.86	22.08	94.07	23.63	93.21	25.15
8	95.58	20.55	94.84	22.13	94.04	23.68	93.18	25.20
10	95.56	20.60	94.81	22.18	94.01	23.73	93.16	25.25
12	95.53	20.66	94.79	22.23	93.98	23.78	93.13	25.30
14	95.51	20.71	94.76	22.28	93.95	23.83	93.10	25.35
16	95.49	20.76	94.73	22.34	93.93	23.88	93.07	25.40
18	95.46	20.81	94.71	22.39	93.90	23.93	93.04	25.45
20	95.44	20.87	94.68	22.44	93.87	23.99	93.01	25.50
22	95.41	20.92	94.66	22.49	93.84	24.04	92.98	25.55
24	95.39	20.97	94.63	22.54	93.81	24.09	92.95	25.60
26	95.36	21.03	94.60	22.60	93.79	24.14	92.92	25.65
28	95.34	21.08	94.58	22.65	93.76	24.19	92.89	25.70
30	95.32	21.13	94.55	22.70	93.73	24.24	92.86	25.75
32	95.29	21.18	94.52	22.75	93.70	24.29	92.83	25.80
34	95.27	21.24	94.50	22.80	93.67	24.34	92.80	25.85
36	95.24	21.29	94.47	22.85	93.65	24.39	92.77	25.90
38	95.22	21.34	94.44	22.91	93.62	24.44	92.74	25.95
40	95.19	21.39	94.42	22.96	93.59	24.49	92.71	26.00
42	95.17	21.45	94.39	23.01	93.56	24.55	92.68	26.05
44	95.14	21.50	94.36	23.06	93.53	24.60	92.65	26.10
46	95.12	21.55	94.34	23.11	93.50	24.65	92.62	26.15
48	95.09	21.60	94.31	23.16	93.47	24.70	92.59	26.20
50	95.07	21.66	94.28	23.22	93.45	24.75	92.56	26.25
52	95.04	21.71	94.26	23.27	93.42	24.80	92.53	26.30
54	95.02	21.76	94.23	23.32	93.39	24.85	92.49	26.35
56	94.99	21.81	94.20	23.37	93.36	24.90	92.46	26.40
58	94.97	21.87	94.17	23.42	93.33	24.95	92.43	26.45
60	94.94	21.92	94.15	23.47	93.90	25.00	92.40	26.50
C = 0.75	0.73	0.16	0.73	0.17	0.73	0.19	0.72	0.20
C = 1.00	0.98	0.22	0.97	0.23	0.97	0.25	0.96	0.27
C = 1.25	1.22	0.27	1.21	0.29	1.21	0.31	1.20	0.34

Minutes	16°		17°		18°		19°	
	Hor. dist.	Diff. elev.	Hor. dist.	Diff. elev.	Hor. dist.	Diff. elev.	Hor. dist.	Diff. elev.
0	92.40	26.50	91.45	27.96	90.45	29.39	89.40	30.78
2	92.37	26.55	91.42	28.01	90.42	29.44	89.36	30.83
4	92.34	26.59	91.39	28.06	90.38	29.48	89.33	30.87
6	92.31	26.64	91.35	28.10	90.35	29.53	89.29	30.92
8	92.28	26.69	91.32	28.15	90.31	29.58	89.26	30.97
10	92.25	26.74	91.29	28.20	90.28	29.62	89.22	31.01
12	92.22	26.79	91.26	28.25	90.24	29.67	89.18	31.06
14	92.19	26.84	91.22	28.30	90.21	29.72	89.15	31.10
16	92.15	26.89	91.19	28.34	90.18	29.76	89.11	31.15
18	92.12	26.94	91.16	28.39	90.14	29.81	89.08	31.19
20	92.09	26.99	91.12	28.44	90.11	29.86	89.04	31.24
22	92.06	27.04	91.09	28.49	90.07	29.90	89.00	31.28
24	92.03	27.09	91.06	28.54	90.04	29.95	88.96	31.33
26	92.00	27.13	91.02	28.58	90.00	30.00	88.93	31.38
28	91.97	27.18	90.99	28.63	89.97	30.04	88.89	31.42
30	91.93	27.23	90.96	28.68	89.93	30.09	88.86	31.47
32	91.90	27.28	90.92	28.73	89.90	30.14	88.82	31.51
34	91.87	27.33	90.89	28.77	89.86	30.19	88.78	31.56
36	91.84	27.38	90.86	28.82	89.83	30.23	88.75	31.60
38	91.81	27.43	90.82	28.87	89.79	30.28	88.71	31.65
40	91.77	27.48	90.79	28.92	89.76	30.32	88.67	31.69
42	91.74	27.52	90.76	28.96	89.72	30.37	88.64	31.74
44	91.71	27.57	90.72	29.01	89.69	30.41	88.60	31.78
46	91.68	27.62	90.69	29.06	89.65	30.46	88.56	31.83
48	91.65	27.67	90.66	29.11	89.61	30.51	88.53	31.87
50	91.61	27.72	90.62	29.15	89.58	30.55	88.49	31.92
52	91.58	27.77	90.59	29.20	89.54	30.60	88.45	31.96
54	91.55	27.81	90.55	29.25	89.51	30.65	88.41	32.01
56	91.52	27.86	90.52	29.30	89.47	30.69	88.38	32.05
58	91.48	27.91	90.48	29.34	89.44	30.74	88.34	32.09
60	91.45	27.96	90.45	29.39	89.40	30.78	88.30	32.14
C = 0.75	0.72	0.21	0.72	0.23	0.71	0.24	0.71	0.25
C = 1.00	0.96	0.28	0.95	0.30	0.95	0.32	0.94	0.33
C = 1.25	1.20	0.35	1.19	0.38	1.19	0.40	1.18	0.42

Minutes	20°		21°		22°		23°	
	Hor. dist.	Diff. elev.	Hor. dist.	Diff. elev.	Hor. dist.	Diff. elev.	Hor. dist.	Diff. elev.
0........	88.30	32.14	87.16	33.46	85.97	34.73	84.73	35.97
2........	88.26	32.18	87.12	33.50	85.93	34.77	84.69	36.01
4........	88.23	32.23	87.08	33.54	85.89	34.82	84.65	36.05
6........	88.19	32.27	87.04	33.59	85.85	34.86	84.61	36.09
8........	88.15	32.32	87.00	33.63	85.80	34.90	84.57	36.13
10........	88.11	32.36	86.96	33.67	85.76	34.94	84.52	36.17
12........	88.08	32.41	86.92	33.72	85.72	34.98	84.48	36.21
14........	88.04	32.45	86.88	33.76	85.68	35.02	84.44	36.25
16........	88.00	32.49	86.84	33.80	85.64	35.07	84.40	36.29
18........	87.96	32.54	86.80	33.84	85.60	35.11	84.35	36.33
20........	87.93	32.58	86.77	33.89	85.56	35.15	84.31	36.37
22........	87.89	32.63	86.73	33.93	85.52	35.19	84.27	36.41
24........	87.85	32.67	86.69	33.97	85.48	35.23	84.23	36.45
26........	87.81	32.72	86.65	34.01	85.44	35.27	84.18	36.49
28........	87.77	32.76	86.61	34.06	85.40	35.31	84.14	36.53
30........	87.74	32.80	86.57	34.10	85.36	35.36	84.10	36.57
32........	87.70	32.85	86.53	34.14	85.31	35.40	84.06	36.61
34........	87.66	32.89	86.49	34.18	85.27	35.44	84.01	36.65
36........	87.62	32.93	86.45	34.23	85.23	35.48	83.97	36.69
38........	87.58	32.98	86.41	34.27	85.19	35.52	83.93	36.73
40........	87.54	33.02	86.37	34.31	85.15	35.56	83.89	36.77
42........	87.51	33.07	86.33	34.35	85.11	35.60	83.84	36.80
44........	87.47	33.11	86.29	34.40	85.07	35.64	83.80	36.84
46........	87.43	33.15	86.25	34.44	85.02	35.68	83.76	36.88
48........	87.39	33.20	86.21	34.48	84.98	35.72	83.72	36.92
50........	87.35	33.24	86.17	34.52	84.94	35.76	83.67	36.96
52........	87.31	33.28	86.13	34.57	84.90	35.80	83.63	37.00
54........	87.27	33.33	86.09	34.61	84.86	35.85	83.59	37.04
56........	87.24	33.37	86.05	34.65	84.82	35.89	83.54	37.08
58........	87.20	33.41	86.01	34.69	84.77	35.93	83.50	37.12
60........	87.16	33.46	85.97	34.73	84.73	35.97	83.46	37.16
C=0.75...	0.70	0.26	0.70	0.27	0.69	0.29	0.69	0.30
C=1.00...	0.94	0.35	0.93	0.37	0.92	0.38	0.92	0.40
C=1.25...	1.17	0.44	1.16	0.46	1.15	0.48	1.15	0.50

Minutes	24°		25°		26°		27°	
	Hor. dist.	Diff. elev.	Hor. dist.	Diff. elev.	Hor. dist.	Diff. elev.	Hor. dist.	Diff. elev.
0	83.46	37.16	82.14	38.30	80.78	39.40	79.39	40.45
2	83.41	37.20	82.09	38.34	80.74	39.44	79.34	40.49
4	83.37	37.23	82.05	38.38	80.69	39.47	79.30	40.52
6	83.33	37.27	82.01	38.41	80.65	39.51	79.25	40.55
8	83.28	37.31	81.96	38.45	80.60	39.54	79.20	40.59
10	83.24	37.35	81.92	38.49	80.55	39.58	79.15	40.62
12	83.20	37.39	81.87	38.53	80.51	39.61	79.11	40.66
14	83.15	37.43	81.83	38.56	80.46	39.65	79.06	40.69
16	83.11	37.47	81.78	38.60	80.41	39.69	79.01	40.72
18	83.07	37.51	81.74	38.64	80.37	39.72	78.96	40.76
20	83.02	37.54	81.69	38.67	80.32	39.76	78.92	40.79
22	82.98	37.58	81.65	38.71	80.28	39.79	78.87	40.82
24	82.93	37.62	81.60	38.75	80.23	39.83	78.82	40.86
26	82.89	37.66	81.56	38.78	80.18	39.86	78.77	40.89
28	82.85	37.70	81.51	38.82	80.14	39.90	78.73	40.92
30	82.80	37.74	81.47	38.86	80.09	39.93	78.68	40.96
32	82.76	37.77	81.42	38.89	80.04	39.97	78.63	40.99
34	82.72	37.81	81.38	38.93	80.00	40.00	78.58	41.02
36	82.67	37.85	81.33	38.97	79.95	40.04	78.54	41.06
38	82.63	37.89	81.28	39.00	79.90	40.07	78.49	41.09
40	82.58	37.93	81.24	39.04	79.86	40.11	78.44	41.12
42	82.54	37.96	81.19	39.08	79.81	40.14	78.39	41.16
44	82.49	38.00	81.15	39.11	79.76	40.18	78.34	41.19
46	82.45	38.04	81.10	39.15	79.72	40.21	78.30	41.22
48	82.41	38.08	81.06	39.18	79.67	40.24	78.25	41.26
50	82.36	38.11	81.01	39.22	79.62	40.28	78.20	41.29
52	82.32	38.15	80.97	39.26	79.58	40.31	78.15	41.32
54	82.27	38.19	80.92	39.29	79.53	40.35	78.10	41.35
56	82.23	38.23	80.87	39.33	79.48	40.38	78.06	41.39
58	82.18	38.26	80.83	39.36	79.44	40.42	78.01	41.42
60	82.14	38.30	80.78	39.40	79.39	40.45	77.96	41.45
C = 0.75	0.68	0.31	0.68	0.32	0.67	0.33	0.66	0.35
C = 1.00	0.91	0.41	0.90	0.43	0.89	0.45	0.89	0.46
C = 1.25	1.14	0.52	1.13	0.54	1.12	0.56	1.11	0.58

Minutes	28°		29°		30°	
	Hor. dist.	Diff. elev.	Hor. dist.	Diff. elev.	Hor. dist.	Diff. elev.
0	77.96	41.45	76.50	42.40	75.00	43.30
2	77.91	41.48	76.45	42.43	74.95	43.33
4	77.86	41.52	76.40	42.46	74.90	43.36
6	77.81	41.55	76.35	42.49	74.85	43.39
8	77.77	41.58	76.30	42.53	74.80	43.42
10	77.72	41.61	76.25	42.56	74.75	43.45
12	77.67	41.65	76.20	42.59	74.70	43.47
14	77.62	41.68	76.15	42.62	74.65	43.50
16	77.57	41.71	76.10	42.65	74.60	43.53
18	77.52	41.74	76.05	42.68	74.55	43.56
20	77.48	41.77	76.00	42.71	74.49	43.59
22	77.42	41.81	75.95	42.74	74.44	43.62
24	77.38	41.84	75.90	42.77	74.39	43.65
26	77.33	41.87	75.85	42.80	74.34	43.67
28	77.28	41.90	75.80	42.83	74.29	43.70
30	77.23	41.93	75.75	42.86	74.24	43.73
32	77.18	41.97	75.70	42.89	74.19	43.76
34	77.13	42.00	75.65	42.92	74.14	43.79
36	77.09	42.03	75.60	42.95	74.09	43.82
38	77.04	42.06	75.55	42.98	74.04	43.84
40	76.99	42.09	75.50	43.01	73.99	43.87
42	76.94	42.12	75.45	43.04	73.93	43.90
44	76.89	42.15	75.40	43.07	73.88	43.93
46	76.84	42.19	75.35	43.10	73.83	43.95
48	76.79	42.22	75.30	43.13	73.78	43.98
50	76.74	42.25	75.25	43.16	73.73	44.01
52	76.69	42.28	75.20	43.18	73.68	44.04
54	76.64	42.31	75.15	43.21	73.63	44.07
56	76.59	42.34	75.10	43.24	73.58	44.09
58	76.55	42.37	75.05	43.27	73.52	44.12
60	76.50	42.40	75.00	43.30	73.47	44.15
C = 0.75	0.66	0.36	0.65	0.37	0.65	0.38
C = 1.00	0.88	0.48	0.87	0.49	0.86	0.51
C = 1.25	1.10	0.60	1.09	0.62	1.08	0.64

D

BASIC PROGRAM FOR STADIA REDUCTION

```
50 PRINT"***********************************************************"
100 PRINT""
101 PRINT""
102 PRINT""
110 PRINT "This program assumes a 1:100 instrument stadia ratio"
111 PRINT"and an instrument constant (C) of 0."
200 PRINT"What is the stadia interval for this observation?"
300 INPUT SI
400 PRINT"What is the vertical angle? NOTE: Angle must be entered"
401 PRINT"as a decimal degree. If the angle is below the horizontal"
402 PRINT"plane, enter a -in front of the angle."
500 INPUT A
550 PRINT"What is the instrument height (HI)?"
575 INPUT C
600 PRINT"What is the elevation of the benchmark under the instrument?"
750 INPUT D
800 PRINT"What is the rod reading at the center crosshair?"
850 INPUT RR
900 HD=(100*SI)*((COS(A*3.14156/180))^2)
950 PRINT""
951 PRINT""
952 PRINT"***********************************************************"
1000 PRINT"The horizontal distance is ";HD ;"feet."
1010 DE=(100*SI)*((SIN((2*A)*3.14156/180))/2)
1020 PRINT"The difference in elevation at HI is ";DE ;"feet."
1030 RE=D+C+DE-RR
1031 PRINT"The elevation at the rod position is ";RE ;"feet."
1032 GOTO 50
```

E

ISO 4463

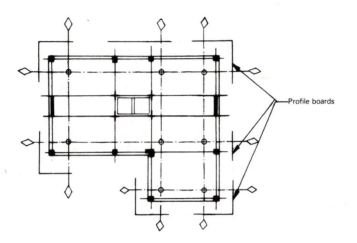

**Figure 5 — Example of how profile boards might be placed outside
a proposed building**

14.3 Marking

Primary points should be established in good time before determining their position to allow for settlement. Whenever possible, primary points should be placed outside the actual working zone on the site.

Where possible a sufficient number of elevated primary points around the site should be established in order to avoid obstructed back sights during the construction (see also 13.4).

14.4 Measuring-in of the primary system

Where the primary system forms a network it has to be observed by measuring sufficient distances and angles to have a redundant number of observations and then be adjusted by the method of least squares. In this way all aspects of quality control are fulfilled.

14.4.1 Distance measurement

There are two alternatives:

a) Using an EDM

Distances to be measured by EDM between primary points should normally be greater than 30 m and measured from each end. The systematic errors of the instrument should be taken into consideration. The instrument should be checked regularly according to ISO 8322-8 or ISO 8322-9 (see also 13.2).

b) Using a steel tape

All distances should be measured at least twice. The values measured should be corrected for temperature, sag, slope and tension. A tension device shall be used with the tape.

Distances to be measured should not be greater than twice the length of the measuring tape to be used. The characteristics of the tape, for example concerning graduation

accuracy and reference temperature, need to be known. If possible, measuring tapes conforming to the OIML recommendations or national standards shall be used (see also 13.2).

14.4.2 Angular measurement

Angles should be measured with a theodolite reading directly to 1 mgon (10'') or better. The measurements should be made in at least two sets. Each set is formed by two observations, one on each face of the instrument. When targets are not fixed as permanent marks they should be mounted on tripods and constrained centring should be applied. Targets shall be well defined (see also 13.2).

14.5 Acceptance criteria for the position of primary points

The primary system should be subjected to two stages of acceptance, namely:

— Stage one, comparing the measured distances and angles with those derived from the adjusted coordinates; and/or

— Stage two, comparing distances and angles derived from given coordinates with those determined by compliance measurement. This can for example be the case on sites where only a coordinate register is available.

14.5.1 Stage one

The relationship between the measured distances and angles and those derived from the adjusted coordinates is examined. The differences shall not exceed the following permitted deviations:

For distances: $\pm 0{,}75\sqrt{L}$
   with a minimum of 4 mm

Courtesy BSI Standards. Extracts from BS 5964: Part 1: 1990 can be obtained from national standards bodies.

For angles:

in degrees: $\pm \dfrac{0,045}{\sqrt{L}} \left(\text{or} \pm \dfrac{2'\,42''}{\sqrt{L}} \right)$

in gon: $\pm \dfrac{0,05}{\sqrt{L}}$

or as offset: $\pm 0,75 \sqrt{L}$ mm

where L is the distance in metres between the primary points concerned; in the case of angles, the shorter of the two distances defining the angle.

14.5.2 Stage two

The relationship between distances and angles derived from given coordinates and subsequently measured distances and angles is examined. The differences shall not exceed the following permitted deviations:

For distances: $\pm 1,5 \sqrt{L}$
with a minimum of 8 mm

For angles:

in degrees: $\pm \dfrac{0,09}{\sqrt{L}} \left(\text{or} \pm \dfrac{5'\,24''}{\sqrt{L}} \right)$

in gon: $\pm \dfrac{0,1}{\sqrt{L}}$

or as offset: $\pm 1,5 \sqrt{L}$ mm

where L is the distance in metres between the primary points concerned; in the case of angles, the shorter of the two distances defining the angle.

For compliance measurements of the primary points, 14.4 applies but a different operator with different equipment of the same accuracy class should be used.

In the case of supplementary primary points which according to 14.2 have been determined by redundant observations (for example point 109 in figure 4), two or more pairs of coordinates will be obtained. Then the mean value of the coordinates should be chosen, provided that the maximum differences between the coordinates fulfil the above recommended criteria.

14.6 Consequences of non-compliance

Before rejecting the primary net, the integrity of all points should be checked. All suspect items (distances and angles) should be re-measured. If compliance has still not been achieved, refer back to the surveyor who carried out the primary scheme.

15 Secondary system

15.1 Introduction

This clause specifies the procedure for setting out and compliance measurement of secondary points, and the acceptance criteria with regard to the internal position of points in the secondary system and between primary points (according to clause 14) and secondary points (see also clause 5).

In this case the coordinates of the primary system are accepted as true and their inaccuracies need not be taken into consideration.

15.2 Application

Secondary points are used for setting-out position points indicating the details of one or more buildings.

The setting-out of secondary points is done by using either

— the primary system as shown in figure 6; or

— previously set-out secondary points (as shown in figure 7).

Previously set-out secondary points should be checked, according to clause 15, prior to any further setting-out.

All the secondary points taken together form the secondary system of the site. Lines through secondary points are called secondary lines.

15.3 Marking

See 13.4 and figure 5.

15.4 Setting-out of secondary points

The setting-out of secondary points should be carried out with redundant measurements and in such a way as to allow for cross-checking.

15.4.1 Distance measurement

There are two alternatives:

a) Using an EDM

Distances to be measured by EDM between primary points and secondary points and between secondary points should normally be greater than 30 m and preferably be measured from each end. However, where it can be demonstrated that a particular instrument has a high accuracy at shorter distances, lengths less than 30 m are acceptable. The systematic errors of the instrument should be taken into consideration. The instrument should be checked regularly according to ISO 8322-8 or ISO 8322-9 (see also 13.2).

b) Using a tape

All distances should be measured at least twice, preferably in opposite directions. The values measured should be corrected for temperature, sag, slope and tension. A tension device shall be used with the tape.

Distances to be measured should not be greater than twice the length of the measuring tape to be used. The characteristics of the tape, for example concerning graduation accuracy and reference temperature, need to be known. If possible, measuring tapes conforming to OIML recommendations or national standards should be used (see also 13.2).

15.4.2 Angular measurement

Angles should be measured and set-out with a theodolite reading directly to 10 mgon (1') or better. The measurements shall be made in at least one set. A set is formed by two observations, one on each face of the instrument.

F

SUGGESTED COLOR CODES FOR FIELD MARKERS

Property corners	fluorescent orange
Jobsite control points	fluorescent orange
Rough grading	red
Finish grade	blue
Sewer lines, manholes, and connections	light green
Storm drain lines	light orange
Storm drain connections and basins	orange
Curb and gutter	yellow
Water lines	blue
Fire protection lines and hydrants	blue
Communication lines	dark green/orange
Gas lines and connections	yellow
Electrical lines	red
Building corners and control lines	white
Wall lines	orange
Excavation	white